LEE COUNTY LIBRARY
107 Hawkins Ave.
Sanford, NC 27330

LEE COUNTY LIBRARY
107 Hawkins Ave.
Sanford, NC 27330

Solar Independent Utility Systems Manual

(A Greener Way of Living) Dedicated To: The cause of a moneyless society and to all who want to save our planet!

Kyle William Loshure

AuthorHouse™
1663 Liberty Drive
Bloomington, IN 47403
www.authorhouse.com
Phone: 1-800-839-8640

©2011 Kyle William Loshure. All rights reserved.

No part of this book may be reproduced, stored in a retrieval system, or transmitted by any means without the written permission of the author.

First published by AuthorHouse   3/2/2011

ISBN: 978-1-4567-3986-7 (sc)
ISBN: 978-1-4567-3985-0 (hc)
ISBN: 978-1-4567-3984-3 (e)

Library of Congress Control Number: 2011902702

Printed in the United States of America

Any people depicted in stock imagery provided by Thinkstock are models, and such images are being used for illustrative purposes only. Certain stock imagery © Thinkstock.

This book is printed on acid-free paper.

The views expressed in this work are solely those of the author and do not necessarily reflect the views of the publisher, and the publisher hereby disclaims any responsibility for them.

Works Cited

Cheney, Margaret. *TESLA MAN OUT OF TIME*. First Touchstone Edition 2001 ed. New York: Simon&Schuster, 1981. Print.

Dukas, Helen, and Banesh Hoffmann. *Albert Einstein Creator and Rebel*. New York: Viking,Inc., 1972. Print.

Isaacson, Walter. *Benjamin Franklin AN AMERICAN LIFE*. New York: Simon&Schuster, 2003. Print.

Baldwin, Neil. *EDISON INVENTING THE CENTURY*. New York: VosBurgh's Orchestration Service, 1995. Print.

Pre-face

This is a manual to all that are totally dedicated to having a solar independent utility system. Meaning 100% of your utilities comes from your surroundings. This is not a whim, idea, hypothesis, or even a dream. Over 30 years of interests have been spent perfecting, inventing, and pursuing the way to becoming utility independent. Starting with solar power and then, adding a dependable refrigerator that never fails by power outages that so many families have been affected by.

Pursuing as well as perfecting, the ability to make it year to year with solar power with a backup generator for the snowiest, shortest days has been realized. The ability to have endless hot water without any utility company has also been realized. Cheap, free, and green was the constant motivator. Dirty expensive water drove it into expanding the independence into water harvesting. Using rainwater and spring water this idea was realized.

Having the ability to totally rely on yourself for all of your household needs will make you feel more secure. This manual will give you the ability to end the bills that will put hard earned money back into your pocket every month for the rest of your life! The child hood dream was made to come true, so it has to be shared with the rest of the world to help make the planet a cleaner and a more stable environment for all who live.

This manual is about changing the undependable and pollution aspect of utility companies as well as their taxing way on the people all over the world. Included in this is information is Einstein, Tesla, Edison, and Franklin. This gives a background on how our electric system we depend on so much today came into existence. The brief overview of their biographies focuses on their contributions to our society. Many of their views on the future outlook of what was to come is included, as well as the best ways to achieve it. Their concern was for the environment as well as the average American. Interesting how Edison for example knew the importance of electric vehicles, one hundred years ago! How relevant is that today as war is raging half way around the world over fossil fuels? Their insight on our past, present, and future is simply amazing.

The passion to be an inventor at such an early age has led to achieve goals far beyond what ever could be imagined thirty years ago but, the mindset was there.

This manual shows what works as well as what doesn't when it comes to solar power. The manual shows very simply how to build a self sustaining system that is dependable and proven from year to year without fail. Revealed as well, is the very uncommon knowledge that will make the difference whether maintenance is required or not. Detailed knowledge is made clear so that frustrating mistakes made by others with good intentions can be corrected. Mistakes made that affect the efficiency so much, that you know your solar system should be producing more, can be fixed. This manual has numerous explanations of reasons of failure as well as to why.

The bad habit of pollution from well intended electric companies all over the world is brought into the light clearly for the inflation, pollution, and unreliability. Effects that the electric companies have had are revealed that affect humanity from freezing homes, as well as roasting homes. Climate change is directly linked to their pollution habits. The electric companies have also caused issues with businesses, and spoiling hundreds of millions of pounds of food. The manual as well has also revealed the hold ups keeping renewable electricity from coming online such as solar power. The electric companies' requirements of a $500.00 minimum production contract a month as well as only giving 50 cents on the dollar credit for renewable power is unfair. Requiring an insurance contract listing them as the beneficiary is just plain ridiculous. You must meet all these requirements to be able to send green electricity onto the grid with some electric companies today. The most shocking issues are that during the sunrise as well as the sunset, the red sky revealing the mercury in the air from burning all that coal contributing to climate change as well as contributing to many health risks.

This has been written to be a manual on how to become 100% totally solar utility independent. Having a solar independent utility system will give you the stability you expect from the utilities without any of the yearly bills that are incurred as well as their pollution and undependability. This manual is extraordinary informative as well as a very good easy educational read.

A note from the editor: As a young teenager in high school I really had passion for writing and spelling. Kyle decided that he wanted to write a manual so I told him I would write as he gave me the information. When I was 20 years old I wrote a poem and got it published in a book with the national poetry society. I also won an award for it so I knew writing this manual was very possible. I have enjoyed every moment of this adventure along the way learning about solar technology and all the possibilities that go along with it.

Amber Nicole Smith-Loshure

Contents

Works Cited	v
Pre-face	vii
Chapter 1) Refrigeration	1
Chapter 2) Water Heating	3
Chapter 3) Electricity	5
Chapter 4) Heating – Vent Free Radiant Heat and Wood Heating	7
Chapter 5) Solar Panels	9
Chapter 6) Battery Banks	11
Chapter 7) Inverters	15
Chapter 8) DC Lighting or Direct Current Lighting	17
Chapter 9) AC Lighting or Alternating Current Lighting	19
Chapter 10) DC Power Supply or Direct Current Power Supply	21
Chapter 11) AC Power Supply or Alternating Current Power Supply	23
Chapter 12) Wire Sizing	25
Chapter 13) DC Water Pumping	27
Chapter 14) Water Filtration	29
Chapter 15) Rain Collection System	31
Chapter 16) Spring Water System	33
Chapter 17) Water Storage System	35
Chapter 18) Water Recycling System	37
Chapter 19) Cooking Methods	39
Chapter 20) Fuses AC and DC	41
Chapter 21) Water Conservation System	43

Chapter 22) Photovoltaic Solar Array Positioning/Sizing/Placement	45
Chapter 23) Grid-Tie Photovoltaic System	47
Chapter 24) Sending Your Power Back to the Grid.	49
Chapter 25) Parallel Photovoltaic System	51
Chapter 26) Standalone Photovoltaic System	53
Chapter 27) Hydrogen Production	55
Chapter 28) Propane Storage	57
Chapter 29) Radiant Heat	59
Chapter 30) Geothermal Cooling	61
Chapter 31) Air Conditioning	63
Chapter 32) Insulation	65
Chapter 33) Lightning Harvesting	67
Chapter 34) Conservation	69
Chapter 35) Construction	71
Chapter 36) Powering Electric Vehicles from your Solar System	73
Chapter 37) Alternative Fuels For Vehicles	75
Chapter 38) Water PSI	77
Chapter 39) The Benefits of Going Green	79
Chapter 40) Power Outages (Grid)	81
Chapter 41) Tesla Technology	83
Chapter 42 Franklin	85
Chapter 43) Einstein	87
Chapter 44) Thomas Edison	89
Chapter 45) Watts	93
Chapter 46) Volts	95
Chapter 47) Amps	97
Chapter 48) Megahertz	99
Chapter 49) Sine Waves	101

Chapter 50) Modified Sine Waves	103
Chapter 51) Square (Block) Sine Waves	105
Chapter 52) Pure Sine Waves	107
FAQ'S	109
Glossary	115
Index	153

Chapter 1) Refrigeration

The best experiences I've had with refrigerators and the most long lasting, cost effective models are the LP gas refrigerators (propane gas model refrigerators). They can run off of DC current that does not require an inverter and they are very long lasting. The reliability is outstanding from year to year. They run off of a mixture of ammonia and salt. The process is much like a regular refrigerator except it doesn't have chemical refrigerant that is harmful to the environment. They are highly reliable in the fact that they run off of 12 volt batteries that are recharged by solar panels and have been proven to operate year to year without fail. With proper battery maintenance and sufficient propane storage it has been the most reliable form of refrigeration that I have ever found.

The issues that I have found are easily solved and are the following: 1) Out of propane. 2) Battery simply needs to be changed. If the propane is out simply turn the unit off for 30 seconds, refill, and then turn it back on. If the battery needs to be replaced simply turn the unit off, replace the battery, and then turn the unit back on. The only major but, easily fixed issue happens in the coldest winter season which is that the ammonia/salt mixture can get a small clog of salt. This is easily fixed by tapping the back of the units heating coils to break up the salt clog.

Periodically defrosting the unit is necessary when frost builds up in the freezer section of the cooling unit. The long lasting ability of the refrigerator is the basic equation of propane, battery reserve, and frost maintenance. With proper maintenance previously mentioned you can expect many decades of dependable refrigeration from your propane, solar powered 12 volt refrigerator. Some newer units require no pilot light which saves gas, is safer, and most efficient.

Kyle William Loshure

    The best part about your solar powered propane refrigerator is the fact that you will never again have to worry about undependable inflationary electric companies that have failed our society and have affected hundreds of millions of families over the past century. You should never trust anyone with the safety of your food except yourself. This is why your solar powered propane refrigerator is the number one most important part of your solar independent utility system.

Chapter 2) Water Heating

On demand water heaters are the best most reliable, cost effective, and an extremely efficient way of having hot water for your whole household that does not ever run out. The way it works is, as the water is passing through the water heater, it is heated to as hot as you desire. Gas propane on demand heater is the best heater in my opinion. When operated from a source of propane and water is pumped by a 12 volt water pump powered from solar power you have an independent source of hot water for showers along with all other uses. Having your water heater this way you can have a total utility independent way of producing unlimited hot water.

The maintenance for your on demand heater is simple but, necessary. When using rain water filtered with activated carbon filters, no maintenance is required. However, if using city water or well water, the lime present in the water will in time accumulate in the heating coils. When maintenance is neglected it will eventually over a long time clog your heating tubes. Depending on how severe the lime is in your water determines how often you need to clean your heater. Through the pumping method of water you simply use a few bottles of CLR to equal approximately a half of a gallon to pump into your water heater and let it dissolve the scaling that has accumulated.

The amount of hot water that you wish to have is a simple equation of gallons of water ready for use, pounds of propane ready for use, and the amount of battery storage power ready for use for water pumping. The right balance of these three supplies will enable you to have an independent utility free capability of producing endless hot water.

Other things to take into consideration are the following: The direct vent straight up will keep you from using an electric blower which will use more electricity. This is the best way to vent your on demand propane heater. Going by the specifications of your on demand water heater is very important in the venting of your water heater. Going by the water heater specifications is also important for your water supply and the propane gas line leading to your water heater. The activated charcoal filter is recommended for your most pure healthy water source.

The location of your water heater should be as close as possible to your most used shower. The size of your water heater is determined by how many showers you wish to operate at the same time. A half of a gallon per minute flow rate water heater is adequate for a household with one shower and normal sink operations. A variable temperature control flow adjustment solar water heater measures the temperature of the water coming in to back up a solar water heating system from the sun or a wood burner heating element used in your wood stove to heat the water. A drain back water valve system needs to be in place if the water heater is ever exposed to temperatures below 32 degrees Fahrenheit as the heater is only heating the water while in use. This is the set of valves that allows the water to drain out of your heater while not in use, while on vacation, or when the expected temperatures are below 32 degrees Fahrenheit.

Chapter 3) Electricity

Electricity has become society's most used asset and valuable in so many different ways. The uses stretch from space all the way down to the furthest depths of the sea. When you think about it and try to imagine a day without electricity think about how much your life would be different. From brushing your teeth, to the food you eat, to the car you drive, to all of the electronic gadgets, internet, cell phones, computers, lighting, air conditioning, and heating would be affected. This has been around for less than two centuries but, if we had to go without it for one day just one day how miserable would our lives be?

The national electric grid has forever changed America but, has also made us more vulnerable than we have ever been. Remember just a few years ago when a tree limb knocked out 1/3 of America's power from 9 seconds for up to two months in some areas. I can think of no better simple thing that has happened that shows just how vulnerable we really are. The real possibilities of you being without power in your own home for months or even years is not out of the question. Things from solar flares, cyber attacks, or physical attacks to the grid can do just that. We should go back as responsible people to what our people founded this country on and that was independence. Electricity is no exception to the rule.

By having your own independent way of producing electricity your life will be more stable than you could ever imagine. Making electricity from the sun is the most stable, predictable, and most reliable way of producing electricity. 12 to 16 volt variable solar panels are the most basic solar panels but, they also make 24 volt solar panels that can be wired in series that can produce a 48 volt system. I prefer the 12 to 16 volt system for the fact that you can utilize 12 volt electricity in numerous ways without an inverter that loses power. You can use inverters to convert a 12 volt, 24 volt, also a 48 volt system to 120 volts, 240 volts, or 480 volts. Different inverters have different power losses from 10% to 30% depending on the quality of the inverter.

Storing your power in batteries is necessary to have a standalone system or a parallel system. When you store your electric power from your system in batteries a very significant amount of power is lost. This is necessary to have a standalone system that is grid free. This is also necessary in a parallel system. Grid-tie photovoltaic solar power will not go into the batteries in the future but, directly onto the grid. This will accumulate a credit by a power exchange with the electric company. Electric companies need to be made to do this as they are unwilling and/or unable to do this in most states. By storing your own power anywhere in the world you can have your own independent utility system that is free of water and electric companies. Your solar electric may be used for factory, businesses, or any other household needs.

Chapter 4) Heating – Vent Free Radiant Heat and Wood Heating

Heating is one of the most vitally important things that a household must have. The sources of heat are highly depended on your household location. You must utilize what is best for your particular location in the mindset of cheap, free, and most abundant resources.

I will start off by talking about all of the different free sources of heat that is possible to utilize. My favorite source would be the hot air panels. The hot air panels simply make heat by the sun passing onto the panel powered by a fan that runs off of the sun. This will give you a good source of hot air that is dependable when the sun is out in a sunny location with a 70 degree angle facing south or north in the southern hemisphere . The hot air panel manufacturer will tell you the specifications and what to anticipate from it. This is one way to have a completely free and totally green hot air supply.

Another method of producing hot air with the same concept but, more difficult is instead of hot air you heat a liquid by the same solar panel method. You take that liquid and put it into a large heat collector that is calculated to last several days. Then you take a geothermal loop from below ground that is calculated to have the capacity to last several days and have an automatic thermal adjusting unit that mixes the two in a heat exchanger that gives you the temperature that you desire. You have two methods of utilizing the temperature that you want your house to be at that point.

1.) The traditional coil unit that air passes through going through your central air ducts is replaced with a coil unit that the fluid passes through to heat or cool your house by a blower that can also be powered by the sun. The fluid is also pumped by the power of the sun. 2.) A newer method that takes the temperature through the heat exchanger passing the temperature through radiant heating tubes and these tubes will run throughout your floor that is made of concrete to evenly spread the temperature that will keep your house at that temperature that you chose by the heat coming up from your floor. This method requires only pumps that can be operated from the sun. You have no air flow that carries dust or other unhealthy airborne particles. This method is also a 100% totally green way of heating and cooling your house to your desired temperature.

    I will go on to talk about simpler and cheaper methods to utilize from your location. You might have a source of natural gas below your property that can be tapped for use for all of your household heating needs, as well as a gas generator. This is free but, not green. You may live in the woods and have an abundance of wood from trees that you can harvest. You do this by taking the dead ones first of course or using the biggest trees from your property leaving the small ones to replenish the woods. This is a carbon neutral way of heating for free. I absolutely do not ever recommend electric heat under any circumstance or situation. The power company that you depend on will charge you a small fortune and is not dependable in any emergency situation.

    Heating your house with electric heat, I also do not recommend, from solar. The most important aspect of heating needs to be focused on independent utility systems. The self-reliance along with sustainability for you and your family needs to rely on your independence of heat.

    A very cost effective way that is not necessarily green but, 98% efficient is the vent-free propane radiant heater. A calculated source of propane to fuel your radiant heater can give you a dependable, reliable source of fuel that can be calculated to last for up to five years without fail. A radiant heater has no moving parts to wear out and will operate decade after decade. One other source people commonly use is natural gas. I do not support this unless previously mentioned that you get it from your own property underground. Anytime that you rely on utility companies monthly bills will occur and inflation will happen. Depending on someone else such as utility companies will leave you vulnerable.

Chapter 5) Solar Panels

The placement of your solar panels is the most vital and important part of your independent utility system. You need to think of your solar panels as one gigantic circuit. Bird droppings, leaves, debris, or even the shading of a telephone wire across your panels can reduce your power output by 15%. Shade just three of the solar cells and your power loss is 50%. Your ability to be able to remove snow and keep them clean is vitally important. Treating the solar panels with rain x will help to shed heavy snow, as well as keep dirt and debris off of them. I prefer the yard mounted system for that reason. The angle of your panel depends on how much maintenance and adjustment that you wish to contribute. A 17 degree south facing mount is best, unless you are in the southern hemisphere, then it needs to be facing north if you wish to have a system that requires no adjustment throughout the year. A flat mount is best on the equator. There are charts that are accurate from any location in the world that will show you a precise angle for maximum solar production.

Panels can vary a lot in wattage from as little as 15 watts 12 volt DC, to as much as 165 watts 12 volt DC. This amount can vary a lot from manufacturer to manufacturer. 24 volt solar panels are also available which carry a significant higher wattage around 250 watts. This can be wired in series to create a 48 volt system. Keep in mind when you wire in series you extend the scope of your circuit which means any shading will even more significantly affect power loss.

Some manufacturers offer a dirt resistance glass and most are well protected from sizeable hail. When mounting your solar system a thing such as the weight of the system needs to be accounted for when placing on a roof. A wind load from wind resistance also needs to be taken into consideration. When I last checked 80% of PV is made outside of America with Japan, China, France, Sweden, and many other countries that are producing PV out there. Thin film solar modules typically black in color work much better in cloudy climates but, they only have a 15 to 25 year life expectancy. Monolithic crystal and typically blue in color have a much longer life expectancy of up to 80 plus years without fail with a normal warranty of 25 years.

The weakest link of a solar system always will be the wires. I strongly recommend every connection from the panel to the battery bank always be soldered or at least have an anticorrosion dressing applied. Depending on how much PV you have in your array will determine how big of wires that you need to carry the watts and amps. You can go by the national electric code book that is updated yearly for the most current updated specifications. The amount of PV you wish to use totally depends on the amount of load you plan on connecting to the PV. There are worksheets out there to help you do the math in coming up with the size of your PV system.

Chapter 6) Battery Banks

Think of your battery bank just like a bank account with the amount that you put in it as the amount that you can spend. Your battery bank gives you the ability to power your house without the power company at night time or during cloudy days. Your battery bank is a key element in that as long as you have sufficient power you can operate an inverter that can produce 120 volts or 240 volts that will supply power to your whole household.

The weakest link in your battery bank is the weakest cell in the battery bank. The way this is tested is with a battery acid tester that will show you when you have a dead cell in a battery. This is important because, if you have a dead cell your entire battery bank will only charge up to the level of the weakest cell in the battery bank. That is why it is important to replace the battery that has the dead cell. Water levels need to be monitored and added to with distilled water to level when it is low. The charge controller will help by stopping the charging of the batteries when they reach a 100% charge. I recommend sizing your battery bank large enough that it reaches around 70% to make your batteries last longer. This will also keep you from losing solar power.

Your batteries can be wired in parallel or in series to create a 12 volt, 24 volt, or 48 volt system. Keeping your batteries clean and treated with anticorrosion cream is very important. When corrosion builds up the power flowing into or out of your batteries will be significantly diminished. This in time will stop the current altogether. Your battery bank can make your solar system very cost effective or very inefficient, if not sized in balance with your solar array. Batteries must always be outside or vented due to the fact that as they charge they will omit a small amount of harmful gases.

Those gases should not be breathed in and could potentially be explosive if the fumes build up and are exposed to a spark.

Batteries charge much less efficient in cooler temperatures and hot weather will cause them to use water at a faster rate which needs to be monitored very closely. Your batteries need to have a fuse in three ways. One fuse needs to be between the battery bank and the PV. Another fuse needs to be between the battery bank and the 12 volt, 24 volt, also the 48 volt power demand. One more fuse needs to be between the batteries and the inverter to meet codes. Deep cycle batteries are very good as they have a lot of cold cranking amps in reserve. Gel cell batteries are very good as well. Large 6 volt solar batteries are very useful as they have the largest reserve cold cranking amps. They can be wired in series to produce a 12 volt, 24 volt, or 48 volt system.

The benefits of a 12 volt system in my opinion are the best. There are many ways in which you can use the 12 volt power to operate many items such as water pumps, 12 volt DC lighting, propane refrigerators, and by utilizing 12 volt you do not need an inverter which will cost you around a 10% to 30% power loss. A 12 volt system of batteries also produces the best results with your PV in the fact that your circuit is smaller and less affected by shading.

Nickel-cadmium batteries have advantages such as long life, low maintenance, survivability from deep discharge, low temperature capacity retention, and non critical voltage regulation. The disadvantages of this battery are high cost and limited availability. Flooded lead calcium sealed vent maintenance free are developed as a maintenance free automotive battery. They are intolerant of overcharging excessive water loss and medium to deep discharge cycles. Gelled VRLA batteries have lead calcium grids and the electrolyte is gelled by the addition of a silica based gel. Silica gel will not liquefy at 40 degrees Celsius. Only ultrasonic vibration or extreme discharge can liquefy the gel. Constant-voltage, current-regulated, temperature compensated charging is recommended and there is the valve regulated lead acid battery. Electrolyte is immobilized sometimes called captive or starved electrolyte batteries. The electrolyte can't be replenished it is intolerant of overcharge minimal maintenance is required, it also is spill proof.

There is a lead calcium grid battery. The advantages include improved mechanical strength, low self discharge rate, and lower water loss than lead-antimony grids. The disadvantage is poor deep discharge performance. The batteries are the most critical part of your solar system in a standalone system, or an independent utility system, as well as a grid-tie solar system with a battery backup.

Chapter 7) Inverters

Inverters are the electronic devices that take 12 volts, 24 volts, as well as 48 volts then convert the power up to 120 volts, 240 volts, also 480 volts for any energy need that you might have whether recreational, household, business, or factory. The cost of an inverter greatly varies. The reason is that there are three different kinds of inverters. The square (block) sine wave is the cheapest inverter. This can run some household appliances but, not all of them. This is good for light bulbs. This is not good for motors or computers for example. The modified sine wave is good for most but, not all household appliances. With this inverter I have had success with computers, microwaves, televisions, and motors. The cost is significantly higher. The pure sine wave inverter is the most expensive type but, very good for ultra sensitive electronics digital equipment.

    All inverters are very dependable. You need to keep in mind however the specifications of the product because, some have as low as power loss as 5% to 10%. Some of them have a power loss as high as 20% to 30%. Most inverters, like the solar panels, again are made outside of America. There is an inverter called a grid-tie inverter which is capable of two things (two different types). There is a standalone inverter as well as a grid-tie inverter which is capable of monitoring the power level in your batteries and when your batteries have a full charge it will send any extra power beyond what your household use is directly to the grid. This gives you a virtual power exchange with the electric company and builds a credit by sending power to them. There is another type of inverter which is a grid-tie only with no battery backup that takes the power from your PV and sends it to the grid when the power level that you are using is less than what you are making. The differences between the two are the following: Your (battery backup) grid-tie inverter will operate when the power is out. The grid-tie inverter (without batteries) will not operate when the power company is out even if the sun is up. The amount of power that you receive and are credited for is much higher with the grid-tie (without the batteries) as you have a very significant power loss in charging the batteries.

Your inverter is the key for your independent utility system as without it you will only have DC power. Fuses are required by code between the battery bank and the inverter. The location of your inverter is extremely important. You want your inverter as close as possible to your battery bank as you will experience a very significant drop of your inverters ability if too far away. 2-OT wire is best to help this issue of power loss caused by to small of a wire if you don't have it close. Larger wire expense is well worth it, as a smaller wire will restrict power over a long distance. Wire to small will drop amps and watts due to the power restriction of a small wire. Keep in mind when building your solar system to keep the inverter again as close as possible to the battery bank and spend extra money on larger than necessary wire between the two.

Chapter 8) DC Lighting or Direct Current Lighting

12 volt DC (direct current) lighting in my opinion is the most basic lighting that a person should have to start with because, by utilizing 12 volt DC lighting you can use a 12 volt PV system in its most basic form without any conversions that will cause you power loss. By using 12 volt lighting even under the worst ambient light conditions when your batteries are running very low you will always have basic lighting.

The newest most up to date light bulbs are the LED 12 volt (diode lights) as they use only 1 watt of electricity to operate. They also make a 120 volt 1 watt LED light bulb. These types of lights are very hard to locate and find. They are also very costly compared to the average incandescent bulb. The light that they put off is moderately dim but, the ability to have lighting decade after decade is possible. There is nothing that can compare with it. The LED light only puts off a fraction of heat that the wasteful incandescent bulb does. Therefore, energy is not wasted in the form of heat.

This is a way to illustrate just how much power that you can save. You can operate 100 1 watt 120 volt LED lights for the same energy consumed in 1 incandescent 100 watt 120 volt bulb. Even if you don't have the money to purchase or can't find the LED lights, a basic 12 volt light bulb will work as long as there is any amount of power in your battery bank. The regular 12 volt light bulb does put off significant heat therefore, isn't nearly as cost effective as the LED light bulbs because, of power loss. I prefer the 12 volt lights with this again as you do not have a power loss occurring with the inverter operating. Basic lighting such as a 12 volt lighting system is a key element in a solar independent utility system. The wiring that you use does need to be sized properly according to the national electric code with fuses to every leg of power supply to meet code.

Chapter 9) AC Lighting or Alternating Current Lighting

AC (alternating current) lighting is possible with your solar independent utility system through the use of an inverter that converts your 12 volt, 24 volt, or 48 volt battery bank into useable 120-240 volt AC current also possibly higher. The power loss that you receive from producing the AC current for lighting is significant but, is still totally functional. LED (diodes) in the 120 volt is available and is the newest most economical way of lighting your house. Compact florescent bulbs starting out at 11 watts put off significant more light than the LED light bulb. The incandescent bulbs can also be operated from the AC part of your solar system. If used I recommend the lowest watt available that you can use and as infrequent as possible. Lighting for your AC lights needs only the basic 12-2 wire unless you have numerous incandescent lights hooked up to it.

The wires also need to have a circuit breaker as most normal households have. Any amount of lighting in the form of AC may be accomplished with the right equation of photovoltaic solar panels, battery bank, and a properly sized inverter. You can expect depending on the quality of your inverter a 10% to 30% power loss in the conversion of DC to AC. Also keep in mind for your AC lighting like other phantom power electronic devices such as televisions, microwaves, air conditioners, or AC chargers plugged in also use power just by being plugged in, as well as in the case of the inverter simply being turned on for AC lighting on demand. A reason to use the 12 volt 1 watt led light is that the phantom load from the inverter simply being turned on is very taxing on the battery bank. The 120 volt 1 watt as well as the 3 watt LED's are extraordinarily good in power savings, although LED's again are not very bright, the power savings are huge. The 120 volt 11 watt florescent lights do a very good job while still saving power.

Chapter 10) DC Power Supply or Direct Current Power Supply

DC power stands for direct current. Direct current means the power is continuous 100% of the time. By the power being such, larger wires are required as power restriction from the direct current is much more prevalent than AC power. This restriction is overcome by larger wires that will enable the power to flow freely without any power drop or overheating wires. Undersized wires could cause the wire to overheat. The determined size of the wire is calculated by the distance from the battery bank to the power being used, as well as the amps and watts flowing over it. This will determine the size of the wire that is needed. This information is located in the national electric code book.

Your 12 volt power in my opinion is most useful as there are many items such as lights, radios, some televisions, refrigerators, and water pumps which are needed for a solar independent utility system. DC power can also be delivered in much higher amounts when your batteries are wired in series to create a 24 volt and a 48 volt system. This is most useful when your load is a very high amperage amount. DC power supply is what your solar panels will produce.

Your wire connections are your most important aspect of your DC power supply. Any wire connection should be continuous without any splits. Soldering is what I recommend if a split is needed. Corrosion will be present in time causing power drops, power restrictions, and ultimately a loss of a connection causing failure of charging of the battery bank. This can also cause failure in the utility usage of your DC power. When and if soldering is not an option they do make an anticorrosion cream that can be applied that will last for a long time. The type of wire that should be used is the fine multi-strand copper wire as power flows most freely on the outside of the small strands of wire. This is why a solid copper wire will work but, not as effectively. Grounding of your DC system is necessary because, it will prevent lightning strikes by a continuous solid wire that connects each panel to another by attaching to the frame of each solar panel in the array without any splits of the copper wire. Connect the wire to the ground rod to the depth that is required by your local code. Your DC power supply should also have a ground wire along all connections of power usage to help assist in the fuse if a short should happen.

Chapter 11) AC Power Supply or Alternating Current Power Supply

AC power is alternating current. What this means is that it is not continuous like DC current but, it pulses at 60 megahertz which is 60 pulses per second of current. Since it does this it allows much more power to run over much greater distances without heating up the wire. One thing about AC current is that it is much more likely to shock you and therefore is dangerous if you grip the wire as your muscles will contract then you will not be able to let go which can lead to death. This is true with 120 volts, 240 volts, 480 volts, or much higher in the transmission wires (wires that run over electric poles). The good thing is that fire chances are greatly reduced. This is much harder to operate off of your solar system at night because the inverter that converts DC to AC has a significant power loss even if the inverter that produces AC is simply left turned on without any load.

Utilizing 120 volts, 240 volts, or 480 volts is much easier accomplished while the sun is shining bright. This is also accomplished when you have a backup generator that produces 120 volts, 240 volts, or even higher which is possible with your solar independent utility system. They make auto turn on generators when your AC power simply needs a boost due to power demand or night time use. This is also very handy if you use grid power (electric company) and their system is down. The AC to DC transformer is also useful to charge the battery bank for short times if the electric company is available and a generator is too noisy. This is not recommended for long periods of time as it will defeat the purpose. AC power supply generators can use gasoline, propane, natural gas, or diesel to operate. The amount of fuel supply equals how much power you have as a backup for your solar independent utility system. Another good quality of AC power, corrosion is much less likely as the current is not continuous.

Once electric companies are cooperative in excepting your surplus of your AC power supply from your solar system you will be sending (with a utility tie system) your extra power on to the grid, giving you a credit with the electric company and that power will go to power your neighbors' house for example. The AC power supply that comes from your battery bank is very limited but, totally possible to power anything. The battery bank needs to be sized correctly by calculating all watts along with amps used with all watts and amps stored in your battery bank. Anticorrosion cream is useful if the wire will not be able to have maintenance on it for several decades. With the use of your AC power supply conservation needs to be thought of in how many watts that you are using in each and every use of all appliances.

Chapter 12) Wire Sizing

Wiring is one of the most key elements in building your solar independent utility system. When dealing with AC current your wire can carry the current much further without power loss and the corrosion factor is much more minimal. Your DC current that runs over wire does suffer a significant power loss even over short distances. The reason this is, it is not a pulsating current like 60 megahertz AC power. The direct current or DC quickly encounters a resistance factor that is not noticeable to the sight or touch. Only electronic testers will be able to verify that you have no power loss from point A to point B. When a split is necessary the only way to have a fail proof system is to solder all connections together or anticorrosion cream may be applied which will help. When there is no soldering or anticorrosion cream you should anticipate maintenance on your system which is simply detaching, cleaning, along with reattaching every connection from your solar panels all the way to your battery bank along to your inverter.

One of the biggest factors for success in your wiring is having proper distance which is as minimal as possible between your inverter and your battery bank. A very critical wiring method also must be used because, of corrosion to protect against lighting strikes. A continuous piece of wire (grounding wire) must be used to ground your entire solar system, this wire will connect from solar panel frame to solar panel frame connecting every panel in your PV array that leads to a grounding rod that meets your local codes as this varies throughout the world from soil type to soil type. This is important because, one lightning strike to an ungrounded system will fry every electronic component throughout your system. I can't stress the importance of this enough.

Another type of wire that is much more suitable for high amperage in your DC wiring is the fine hair like strands of wire otherwise known as locomotive wiring. This is a very flexible and extremely thick wire that is best for connecting your inverter to your battery bank also in cases of a very large PV array. One key element of electricity passing over wire is that electricity flows not in a wire but, the electrons move across the outside of the wire (beneath the insulation sheeting of course). When all of the mentioned information above is strictly followed, your solar independent utility system will be as reliable as the sun shining on your solar panels.

Chapter 13) DC Water Pumping

With an independent utility system your DC water pump will be the key to endless water pumping for your shower, sink, toilet, and any other water needs that you might have. I prefer the DC pump. The DC pump can utilize solar power in its raw form even without batteries while the sun is shining. With a battery bank you can size your system to meet all of your water needs year round anywhere on earth. The reason that I prefer DC water pumping is because, there are no inverters needed which have power loss in the conversion of DC to AC. There is also the consumption of power simply being turned on in standby mode from the inverter. DC water pumping is very important as in long periods of low ambient light, your batteries might not be sufficient to operate the inverter for AC water pumping. The flow rate is typically around a half of a gallon per minute but, can be much higher in larger water pumps.

    Your DC water pump will give you the ability to reduce or totally eliminate the need for a water company. By utilizing what is available at your property location for example: The rain that falls on your roof can be collected, filtered, stored, and pumped on demand. This works by the pump in a way that monitors the pressure in your water system (water line) and when the pressure drops below 40 psi your pump will kick on automatically. When you turn on your shower for example your pump will automatically turn on and when you turn your shower off it will turn off automatically. I recommend an activated carbon filter to purify your water on the pressure side of your water system. You need to use the water source with a combination of water sources to achieve enough water harvesting to accommodate any of your water needs.

I prefer rain water as it does not have lime sediment that is found in pond water, spring water, city water, or well water that may need to be used again in combination to meet your water demands. The power supplying your water pump needs to be stable. The best way to achieve this is to have a continuous wire (without splits) from your battery bank to your PV directly and soldered connection to your water pump. Anticorrosion cream may also be applied. This will ensure that your DC water pump will operate without fail. Your DC water pump is going to be a significant source of power consumption in your solar independent utility system.

Chapter 14) Water Filtration

Water filtration can be accomplished in many different ways. One of the most expensive and purest forms is reverse osmosis which is followed up with UV light radiation. One of the cheapest forms that are still highly effective today dates back to the earliest European way in which you pass water through sand, activated charcoal, and limestone rock. This is a very cost effective method it does however get contaminated in a short time and will need to be changed. Some pre-filtered methods are very useful. These types include: Two micron cotton filters which are available at most places followed up by an activated charcoal filter. This process simply passes the water through these filters before it reaches your sink, shower, and so on. Pre-filtering water with gravity is a very good method as water is being collected from the roof. This simply passes through polyester nylon fibers that resist mold and growth.

Another common practice from the longest time ago in the Roman days is simply to let the water stand for several days to let the sediment settle before filtering with other methods mentioned above. Another method which is high tech but, almost unheard of is ozone purification. This process ionizes the water and sterilizes it. There is also the much more common method known simply as chlorinated water. All of the mentioned above will let you achieve safe successful water harvesting from rain water, spring water, well water, lake water, or even pond water. This is a very vital part of your solar independent utility system.

Chapter 15) Rain Collection System

For your solar independent utility system your ability to collect and store water is actually more important than the electricity that you are gathering with your solar panels. The method of collecting rain water is going to vary alot depending on your household. The fundamentals are basically the same. You will use gravity to your benefit to use the flow to force the water through a basic filter that lets large quantities of water flow through. This filter needs to be antigrowth from algae, mold, bacteria, and so on. Then the water will flow into a containment system such as a water cistern. The amount of water that you can store will enable you to make it throughout the year.

    I recommend below ground storage for the basic reason that it keeps the water from freezing in the winter but, that also creates a cleaning issue that can be overcome by a suck pump. During the winter months an inside water storage system is also recommended. The amount of water that you can collect will vary so widely all over the world. You need to calculate your average annual rainfall for your particular area times the square footage available on your roof to be able to determine about what to anticipate. You should double what you think you can collect per 3 months so that you have the ability to overcome droughts and the deep freeze of winter. The water purifications mentioned in the previous chapter will give you the ability to utilize all of the water you collected safely.

    You do not want to use containers that will rust, decay, or degrade overtime. Keeping all insects out of your water is very important as it will contaminate the water in a very short time and make it unusable. Rain water is very important because, the amount of sediment is the lowest among the water gathering methods so it won't deposit lime sediment in your water heating system. Rain water again is very vital to your solar independent utility system.

Chapter 16) Spring Water System

The ability to collect spring water from your property when available will greatly increase the amount of water you have available for all household or other uses. The key to being able to get the most water for the least power is a technique that is very simple in the concept but, a bit hard to implement. When attaching the water pump to the water line that leads to your spring, you must have in place a way to prime (or fill) the water line with water from the pump head to the spring. Any amount of air will significantly handicap the ability of the water pump to prime itself. A basic filter needs to be in place at the collection site. I have had success using only 6 volts of power on a 12 volt water pump to draw hundreds of gallons of fresh spring water.

The water quality will greatly vary around the world and the sediment needs to be filtered out before use by storing in a water collection system that will be further filtered by another pump that will bring it up to usable quality. Do keep in mind that redundancy of the methods of using water off of your property such as rain water, spring water, and well water will be needed to ensure that you do not ever run out. I prefer the 12 volt water pump for collecting your spring water because; with a 12 volt water pump you can use your solar power without power loss due to an inverter that converts it to 120 volts. The 12 volt system is also best for spring water collection because, your battery bank sometimes will not have ample power to operate the inverter, also the 12 volt system will work with even the most discharged battery bank. Your spring water collection system will greatly enhance your solar independent utility system.

Chapter 17) Water Storage System

The water storage system that you choose is greatly going to be dependent on the availability of resources at your disposal and the financial ability to create as big of a system as you anticipate needing. You want to have the ability to recycle the water from your tank to your water pump through your filter and back to your tank again so you can have the ability to purify the water that is in storage as needed. Plastic tanks are typically the best as they resist ice cracking, deterioration, or contamination.

I typically recommend a water storage system of approximately 500 gallons per person of storage to have the ability to overcome the winter freezes and summer droughts. You also need the ability for simple cleaning maintenance to remove sediment that will settle at the bottom of the tanks over time. Your water storage system is also a key element in you solar independent utility system. Having an inside water storage system is important to overcome hard winter freezes.

Chapter 18) Water Recycling System

Water recycling is a very important part of your solar independent utility system for a few reasons. The most important reason is because, of cold temperatures. Recycling the water in your household taking from one end of your water system and supplying it to the other end of your system what you do in effect is send water continuously through your entire water system. Doing this the water inside your house will have a certain amount of heat in the water tank that will spread throughout your house which will keep all of your water lines from freezing up even in the most extreme freezes.

Another aspect that is very important is to be able to take the water that has been standing too long and recycle it through your activated carbon filter. The more times the water passes through the system the more pure it will become. This is done simply by drawing from the water supply through the pump, into the filter, into your water system, and then back to your water supply. Your water supply system, water pump needs to run off of 12 volt DC to avoid power loss caused by the voltage drop from DC to AC as well as the power loss from the inverter simply by being turned on. Running off of 12 volt DC this will give you the most dependable system that will not fail unless your batteries are totally drained. I have had luck with even a 6 volt power supply on a 12 volt pump to save even more power. This is a good system to have in place if you live in cold climate areas or if you tend to let your water inside ready to use supply standing for more than one week. You will find having this system in place extraordinarily useful especially in the hard freeze of winter. This is a very useful part in water conservation and dependability for your solar independent utility system.

Chapter 19) Cooking Methods

There are a wide variety of cooking methods that can be used. Depending on what is most available and cheap in your area. You might have an abundant amount of wood on your property or a natural gas reserve under your property. You need to decide what is the most available and the most cost effective to your location. I personally like propane stoves as you can have an abundant supply of propane that keeps you from being dependent on unreliable gas utility companies or electric companies that in time due to inflation will always cost more. They are also almost never there in periods of true emergencies.

I also like wood stoves to cook off of as they are carbon neutral and can be multi-tasking in your home, heating it while also providing an instant heat source to cook your food. This can also heat your water for your household all at the same time. I do not recommend electric for your cooking methods because, they draw simply too much power from your solar independent utility system. Outdoor summer kitchen should be made to avoid adding heat to the household during the summer if possible. This is the most dependable and cost effective way to make your solar independent utility system to work for you in cooking your food.

Chapter 20) Fuses AC and DC

The fuses between AC and DC power are very different. DC power is typically a burnout type fuse much like the type you see in a car and will need to be replaced whenever it might burn out. The AC fuse is typically located in your breaker box and will trip when the power load exceeds the rated use of the breaker. Typically this will be measured in amps and the DC fuse will typically also be rated in amps. They both are typically located or should be built into your house in very visible areas. There are some very old AC fuses that screw in and when they blow must be replaced just like the DC fuses. When resistance is a concern for your DC fuses they do make a gold plated fuse of various sizes that will resist corrosion (that cause power loss) but, they are much more costly.

The wires coming from your AC along with DC fuses must be rated to handle the current that the fuse will provide on both AC and DC. Fuses should be located in your solar system between your PV array and the battery bank that is rated for your PV array maximum power output. Fuses should also be located between your battery bank and DC loads. There should also be fuses between your battery bank and your inverter. A special fuse switchover to utilize your solar panel power from the AC inverter in parallel with the electric company should be the type that only allows either your PV power on the AC side to be used or the electric company power to be used. This is a three way type of fusible switch that is to be used in a parallel solar system with the electric company that will keep the two power supplies from being connected at once. When the utility grid runs into PV power without a grid-tie system it will damage your inverter. That can be very costly. By doing all of the mentioned above you will meet all national and local electric codes that will keep your system safe from any harm. This will also satisfy all required codes. This is also a very important part of your solar independent utility system.

Chapter 21) Water Conservation System

Water conservation is very important especially if you are going to just utilize rain water, well water, spring water, or have extraordinarily expensive water from water companies that is dirty, impure, contaminated, unsafe to drink, unpredictable when you are going to get it shut off, also if they impose water restrictions involuntarily. Here are some of the methods that I have learned: On demand water heaters located as close as possible to the shower will help to save running water while you are waiting for it to heat up. The 12 volt water pumps that have a 40 psi rating will help to slow the water flow which will also save a lot of water when you are using your faucets.

Very low flow toilets are available that have a number one and a number two button that will greatly reduce the water that you need. Instant shut off switches located on the tap (on the end of the faucet) are as well very useful. Even the location of your shower head to be located above your tub that lets the water fall straight down will allow you to use much less water flow while showering. There are of course the things that we have heard about for years such as shutting off the faucet while brushing your teeth, limiting your showers to what is necessary, staying away from automatic dishwashers, and high flow toilets. There are automatic faucets that turn on and off by a sink censor, there are also the push button faucets that also turn off automatically if you want to be extravagant. All these methods mentioned previously will greatly reduce the amount of water

Kyle William Loshure

that you are now using unnecessarily and will enable success of your solar independent utility system.

Chapter 22) Photovoltaic Solar Array Positioning/ Sizing/Placement

The placement of your PV system should start with the following; a solar path finder should be used in a few positions in your yard and a few positions on possible spots on your roof. The solar path finder will reveal all available sun hours and will also show you all shade locations. You can calculate each spot that you think might be a good area. This will give you an exact calculation of how many hours of sunlight you will receive. This will help you determine in part where the best location will be. You want to position your solar system due south with the angle in Indiana, USA for example 22 degrees facing south. This will greatly vary depending on where you are in the world, for example if you are in the southern hemisphere you would want it facing due north. The flat mounting position would be best if you are close to the equator. The 90 degree facing towards the equator is best if you are in the north or south poles. This will also help shed heavy snowfall amounts. During the summer season a tracking PV array that will follow the sun and the PV will literally go in circles as the sun travels across the sky as the sun does not set in the summer time in the North Pole as well as in the South Pole.

    Another factor that needs to be taken into account into the positioning of your PV is the amount of PV that you have. The largest panels for example more than a half of a dozen do have significant weight to them. Not only the weight of the panels needs to be taken into account if your roof will hold it but, wind loads as well. The wind loads need to be taken into account for example in North Dakota, USA also in Florida, USA because, of hurricanes. The accessibility also needs to be taken into account for example, if you live in a dusty area or a heavy snow area you will need easy access to the panels for purposes of cleaning along with snow removal. What might work best for you is a yard mounted system. My opinion is that I like the yard mounted system best for maintenance and upkeep. The security of your solar panels also needs to be thought of as you don't want kids breaking the panels with rocks or theft might be an issue in your area. The things mentioned above needs to be thought through carefully before positioning your solar array system.

The size of your PV array is totally going to be determined by the amount of power in which you need to produce. This can be calculated by watts, times hours, times days, in comparison to available ambient light as shown in your solar path finder, times the amount of hours, and the amount of PV needs to fit the equation between the two. There is very good information that will show you what you can expect from your particular location in the world. This will also help you determine just how much PV that you will need. I recommend after you think that you know what you need adding additional 30% more PV to ensure the best results and additional electric availability to meet your future needs.

The placement of your solar system can best be decided on the above information mentioned but, ultimately the decision is up to you. This can also be up to the contractor that you hire to install the system. One additional thing that needs to be mentioned is called the solar array tracking system. This allows the PV to follow the sun across the sky, I have heard of a few complications that cause minor issues that are easily fixed by a licensed PV specialist. The solar tracking array can increase your power by around 35% over the course of a day. You might wish to adjust the angle to follow the sun as the seasons change all over the world in a fixed PV system. All of the mentioned information above will help you to achieve the very best results of your solar independent utility system.

Chapter 23) Grid-Tie Photovoltaic System

Grid-tie is the new and upcoming revolution in America's power as well as around the world. The basic ideas behind grid-tie photovoltaic systems are the most simple and fail proof ways of getting the most power out of your system. How this works in a grid-tie only system is that your inverter acts the same as the transformer on your electric pole. This is different in the fact that it receives power from both your solar system and the electric company to make up the short comings of your solar system at night, also in cloudy conditions. When there is an abundance of solar power which exceeds the amount of power that you are utilizing it simply sends the power back onto the grid in a virtual power exchange that will give you a credit when your power surplus from your solar system exceeds your demands over the month.

You can typically expect between a 10% to 20% power-loss in the conversion of DC to AC power through your inverter. With a grid-tie only system (without batteries) your system will monitor the grid and will not operate when the grid is down. You need a battery bank back up that will be used if you wish for your system to operate when the grid is down. The way this works is the first power made in excess of what you are using will go to charge the battery bank first. Then once the battery bank is full (your inverter will monitor this) all surplus power beyond what your household use is will be sent onto the grid in a virtual power exchange over the course of a month. You will have a credit for the month if your usage is less than your power produced from your solar system.

The benefit of this system is that when the grid is down your battery bank will operate your household as long as the power is available in your battery bank. The system will operate continuously with out fail when the sun is out, the grid is down, and the amount of power that you are producing is the same as you are using. When the grid is down with this system and as night sets in, any short fall from your PV array will be made up by your battery bank, as long as there is sufficient power available to operate your inverter. The benefit again in this system is the fact that it will operate when the grid is down versus the grid-tie only system (no battery bank backup) which will not operate when the grid is down. The power production in the grid-tie only system (without batteries) is significantly higher. The grid-tie system with the battery backup is significantly less power production sent to the grid for the reason that a substantial amount of power loss occurs in the process in charging the batteries.

I personally prefer the grid-tie system with a battery backup for the reason that the electric companies are not dependable in times of true emergency and this is another key element in the success of your solar independent utility system. Our society in my opinion needs to legislatively force the unwilling or unable electric companies to accept any amount of power being produced which is sent to the grid instead of requiring a minimal amount of power of $500.00 a month, by contract requirement, for example as some electric companies currently require before they will accept any amount of power. Furthermore all electric companies need to give a dollar for dollar credit instead of a 50% reduction of credit. There are some electric companies that will only credit you half, 50 cents on the dollar, for the power that you send to them. This needs to be changed to a dollar for dollar exchange. They do not need to require insurance carried listing them as the beneficiary and having us paying for it as well.

Chapter 24) Sending Your Power Back to the Grid.

Sending your power back to the grid is best understood by understanding the megahertz. When the power comes in from the grid to your home it comes in at 60 pulses per second (60 megahertz). When sending power onto the grid it is done in ever so slightly higher of an amount so that the power will flow backwards through the transformer back to the grid. The transformer on your electric pole does not need to be changed or even any alterations do not need to be made, it is ready to accept power from the surplus power of your solar system.

    The only thing that needs to be done is to have the electric company place a meter that will run backwards, sometimes a dual metering system may be installed and you are ready to send your power onto the grid. Some electric companies will require a net metering which simply keeps track of the power that you send to them and the power that they send to you. No wires need to be changed from your electric box to your meter or anywhere else. An electrician simply needs to connect your grid-tie inverter from your PV to your electric box. This is just as simple as it sounds. The last time I talked to an electric company they wanted to have a $500.00 minimum production requirement and only give 50 cents on the dollar credit. They wanted a contract signed and insurance instated with them listed as the beneficiary in addition to the other things which are listed above. My opinion is what needs to be done is legislatively get rid of this nonsense and force them to accept the clean green energy of solar power. This will make your solar independent utility system not only more

Kyle William Loshure

productive for you but, also will contribute to a greener society as well as the national electric grid.

Chapter 25) Parallel Photovoltaic System

A parallel photovoltaic system is a solar system that operates primarily from solar energy but, uses the electric utility grid as a backup. Effectively you are using solar energy next to but, not connected to the electric company. This is done for the main reason that electric utility companies have unreasonable quota demands and unnecessary insurance requirements as well as cheating us only giving us 50 cents for every $1.00 of power that is being produced.

A parallel photovoltaic system operates in the following way: You have your battery bank which stores the power. Then you have your PV which produces the power as well as the charge controller that controls the charging and shuts off the power when the batteries are full. Sizing your battery bank large enough will make it possible to keep the charge controller from shutting the charging off and making the most use of your solar power. Then you have your inverter which steps up the power from 12 volts, 24 volts, 48 volts to 120 volts, 240 volts or 480 volts for example. This does not connect in any way to the grid. Therefore you do not need those ridiculous contracts or any of their other unreasonable demands.

How this parallel system is connected is that there is a three way breaker switch which allows you to use either solar power or electric company power to operate your household. This is not as cost effective as a grid-tie only system, but it is however as reliable as a standalone system. This is also as reliable as a system that is grid-tie with a battery backup. I have not yet found a way to make an automatic switch that takes it from solar to grid for example, when the sun goes down or when the battery bank is low. This will be the next part to no manual involvement in utilizing your solar power. This is just a simple switch now that you switch from solar to the grid when necessary. A parallel system is probably going to be the most used as people make their way towards a 100% solar independent utility system.

Chapter 26) Standalone Photovoltaic System

A standalone photovoltaic system is a system that produces solar power that can be used to power your every electrical need all day with limited power at night depending on your battery bank supply. The way this process works is when the sun is shining on your solar array the photovoltaic cells in the solar panels collect the solar power sending it through properly sized wires that are soldered also anticorrosion cream could be applied to every split in the wire (if not soldered) from the solar panels to the batteries with a fuse between the PV and your battery bank. The charge controller needs to be between the batteries and the solar panels to keep from over charging your battery bank. I like to size the battery bank large enough that they never get much above 70% charged as to not waste any solar power.

    The best experience I have had with storing power in a 12 volt system wired in parallel is to connect the positive on one side of the battery bank and the negative on the opposite side of the battery bank. This method of wiring for the batteries needs to be applied in both the charging and the battery bank load. The wire between your battery bank and your inverter must have a fuse. The wire must be properly sized (I strongly recommend the inverter being placed as close as possible to the battery bank to avoid power loss). The 12 volt load from the battery bank to all of your DC loads must also be fused. The wire must also be properly sized to avoid power loss or potential fires.

    The size of the inverter that you have totally depends on your power consumption also your battery bank supply needs to be based on expected kilowatt use over several days or weeks. The type of inverter that you use in your standalone system will depend on the sensitivity of the electronics that you plan to use. With sensitive computers you will need pure sine wave inverters. You will want modified sine waves for televisions and many other similar electronics. Block sine wave inverters are best for very basic uses such as the old incandescent light bulb.

I have found personally that the very best choices of inverters are the modified sine wave inverters for their cost and for their wide range of electronics that they can operate. The dependability of your inverter is extraordinarily reliable when wired properly and the wide range of electronics that it will operate will simply amaze you as it has me for well over a decade. Your standalone PV system is based on the ability of your inverter, your battery bank capacity, and your PV production of power. With a standalone system you are producing, storing, and using all of your own power free from any electric company. These systems can be built absolutely anywhere in the world, so if the sun shines where you are you can to have your own solar independent utility system.

Chapter 27) Hydrogen Production

Believe it or not you can produce hydrogen from your solar system. I have found in the methods that I have tried and succeeded at, it uses a lot of power in the production. The basic concept however is that you want to put together two metal plates on opposite ends of the water that spread the current over the entire plate and pass electricity through the salt in the water with a tad bit of baking soda to stabilize the solution. You will see the water actually boil but, no heating is being produced. The bubbles that are coming up are actually hydrogen split from the water molecule. What comes out in the bubbles is a hydrogen and oxygen flammable air from only water.

The issues that I have found with this are the following: It consumes a whole lot of solar power. The hydrogen that is being produced even pops like a firecracker when lit in small quantities so it is extraordinarily explosive. This is why I haven't found any uses for it yet. The final issue with it is that to liquefy it, it has to be at many times the psi of natural gas or propane. Storing the gas I see as very dangerous. Maybe in the future even the near future other inventers and scientists will find the proper way to utilize the endless supply of hydrogen fuel that is available on our planet that is pollution free. Utilizing hydrogen as its being produced will help to overcome storage issues with hydrogen and may well open endless opportunities with it.

Chapter 28) Propane Storage

Propane storage for your independent utility system as of today is a very important fuel source for your refrigerator, water heater, propane cooking, as well as radiant heating. The storage of your propane should be thought about carefully. The abundance of propane will ensure the long availability between refills but, the hazardous risk goes up with every gallon stored on your property. Common problems that I have seen are if a leak develops in your copper gas lines you can not only lose all the gas in your tank but, potentially could be an explosive hazard that can even destroy your home.

Where you store your propane tank should have some thought in it as with anything that is explosive. Although they have been proven very safe and very reliable over many decades they do play a very key part in your solar independent utility system. One final issue that I have also seen is during floods they tend to float, which can result in issues. Anchoring them down will keep them put for floods as well as possible thefts. Properly sized copper lines must also be followed as well for proper operation of your appliances such as the on demand water heater, LP gas refrigerator, radiant heater, and cooking stove.

Chapter 29) Radiant Heat

Radiant heat is a very excellent source of heat especially when coming in from the cold. This penetrates and goes deep into your body warming you to the bones unlike for example, central heating from a heat pump that the air temperature only comes out marginally warm. There is no better heater which gives an example of this than the old fashioned wood stove. Perhaps the heat that warms you from the leaping flames of a camp fire or a charcoal grills red embers is something that you might have experienced if you have never been around an old fashioned wood stove.

Radiant heat comes in waves instead of fan forced air exactly like the sun does and therefore is the reason why our bodies like it the best. Radiant heat can also come from absorbing the suns heat through hot water collection tubes then regenerated throughout your home again. This can also come from sources such as a propane vent-free radiant heater that is 97% efficient and does an excellent job at taking off the winters chill. The very newest radiant heat technology is a technology which takes radiant heat and distributes it equally through tubes throughout your floors. This also distributes it equally throughout your walls in your home evenly to create the temperature that you want stopping any cold surfaces to the touch throughout your floors and walls.

Using radiant heat technology in this way will give you the most comfortable house that you have ever had. Using the radiant heat from a sunny day will give you the cleanest greenest type of heating that is possible. Propane stored in bulk quantities can give you the most abundant amount of radiant heat. Wood heat when using old dead wood from the forest will give you a carbon neutral source of radiant heat. There are the less dependable, unreliable, inflationary sources that I do not recommend such as electric companies and natural gas companies. These however are very commonly used today. Radiant heat can be stored as just mentioned before such as in large propane tanks and large supplies of firewood. There is the newest technology which is capturing the suns heat through piping that heats the liquid in the tubing and stores it in a properly sized insulated tank for usage when needed. These are the best applications in radiant heat that I have ever found for your solar independent utility system. These systems use the power from the sun to operate all the pumps and fans so it works very well with your solar independent utility system.

Chapter 30) Geothermal Cooling

Geothermal cooling is the most awesome way of cooling your house that when accompanied with the power of solar energy to run the pumps that circulate the water and the fans that circulate the air is the very cleanest method of cooling your house with absolutely zero impact to the environment. How this works is very simple, you have a very large water source which is sized accordingly to accommodate your cooling needs for your house, office, or any business use. The pumps simply move the water through pipes deep in the ground that cool the water in the tubes, as it passes through the ground, then it is recycled through a radiator type setup that air passes through, taking the coolness from the ground, and spreading it by fan forced air for your every cooling need.

Some geothermal type systems are basically installed just like wells and can be a little costly to install. Some geothermal systems can also easily be installed very cost effectively when breaking ground for a new house, while the excavation crew is already digging basements, and while all of the equipment is already there. This is simply done by using large sections of your land at the 55 degree temperature line that varies throughout the world and has enough tubing to meet the capacity that will accommodate the amount of cooling needed for your particular installation. Geothermal technicians can help assist you with these calculations.

Technology is also there that when accompanying geothermal technology with radiant heat technology throughout tubes located in your floors and walls will give you the healthiest method of cooling as well as heating your house. This will also not stir up dust mites, dust, pollen, viruses, and other airborne allergens that will greatly affect the quality of health in infants, elderly, and your family. This is the most advanced technology that I have found to make your solar independent utility system the very best that it can be both economically and environmentally.

Chapter 31) Air Conditioning

Air conditioning in respect to the environment is best accomplished with the geothermal cooling method but, if this is not possible other simple fixes can be implemented. You can simply check for air leaks in all of your windows and doors fixing them properly if needed. Adding insulation to your roof can play a very useful part in keeping your home or office more comfortable year round. Reflective roofing material can also be very helpful and useful. More insulation added to your walls if possible is recommended. The type of building material that you use will play a key part to the amount of air conditioning that you will need. There is even a reflective type of wall wrap that is very effective in keeping heat out and air conditioning in.

There are other types of interesting technologies such as automatic setback thermostats that will cool your house during the hours that you program it for. There are high efficiency air conditioners that have very low wattage use but, cool very effectively. Some even have timers that automatically shut the air conditioner off in desired hours which you choose or high efficiency models that automatically choose what temperature to cool your house to going by its own efficiency readings for that particular day.

The main comfort achieved by an air conditioner is lowering the humidity level to 50% or below, making even an 85 degree temperature much more tolerable, as your body can more effectively cool itself in a lower humidity environment. There is also the common sense approach of dressing accordingly with shorts, short sleeves, and shoes that promote breathability of your feet. In order to be able to utilize your solar system to produce power able to operate the smallest air conditioner you will

Kyle William Loshure

need around a 2 kilowatt system to effectively cool your house without the pollution that the electric companies cause. This is the best information that I have found to operate your air conditioner with your solar independent utility system.

Chapter 32) Insulation

Insulation will be a key part in the comfort of your home, office, or factory from season to season as the human race has been striving to create our own comfort level without regard to the elements of our environments. There are a few fundamental facts regarding insulation that will greatly affect your comfort level. Number one is air flow, if you do not have the ability to stop the wind; you will either freeze or roast depending on the season. There is the density of the walls which are made up by not only the building material but, also the amount of insulation inside. Lastly is the ability to stop radiant heat from being transferred in or out of your home as well as your office environment.

With some building materials it is possible to have too much insulation, as this could cause a micro environment, which can cause condensation that can lead to water problems. This is the best information I have to help assist you with your solar independent utility system.

Chapter 33) Lightning Harvesting

Lightning harvesting is a dream that I believe Ben Franklin had. Every since I started making my own electricity from the sun I kept on being drawn back to what endless abundant possibilities there would be if everyone on the planet could start using the supply of lightning for an endless source of power because, lightning hits all around the world at the rate of about 100 times per second. Talking about green technology and zero greenhouse gases how the earth would be changed overnight if this renewable source was utilized as well as solar power.

Our imaginations simply could not wrap around how abundant it is. I only have child like dreams of how it could be possible with the technology that exists today. I don't want to go into any detail of what my thoughts are as to not pollute any inventor's or scientist's concept on how to achieve this. I wish our human race would seriously revisit this idea.

Chapter 34) Conservation

Conservation should be a key part to your solar independent utility system, unless you have unlimited funds to build your solar system, to give you unlimited power for the future. LED lights are the very most economical as they only use 1 watt and above. LED lights do however put off a very dim light. Ten, thirteen, twenty watt florescent bulbs are great as they are up to 10 times more efficient than incandescent bulbs, as well as having a much longer life expectancy. Basically anything that generates heat from electricity whether an incandescent bulb or a furnace will not be a part of your independent utility system, unless you have a very substantial square footage area to collect the PV power, a very large battery bank to store it, as well as unlimited funds to do so.

By not converting DC into AC and utilizing only the DC current the conservation of your solar power will hugely be amplified. By picking out the type of inverter that has the lowest power loss such as a 10% power loss inverter, will help to lower the power loss, in the conversion. Consider using a very small inverter for your small power loads that you use all the time to lower the phantom power loss from your inverter. Eliminating all of your phantom power such as transformers, televisions, microwaves, and many other electronics when plugged in that consume wattage even when not turned on, is one of the most key elements in the conservation of your solar electricity. The type of televisions that you choose to use, you need to keep in mind the wattage that they are consuming as well as computers, fans, air conditioners, and all other electronics.

You can conserve your water in ways such as water shut offs at the end of your kitchen sink and 0.8 gallon high efficiency toilets. On demand water heaters located very close to your shower will also greatly restrict the water that is being lost while waiting for hot water. There is the obvious conservation of the water flow rate (or psi). Other methods of conservation are definitely very important. Wherever you can find ways of saving water or electricity, you must do so. Conserving your battery bank for overnight hour's usage is best done by simply using the AC power when the sun is shining on your PV. Using your high electricity loads such as vacuuming and television hours should be used primarily between 11am to 3pm wherever you are at in the world. This is the peak energy production time and you get your best conservation in using the power as it is produced to avoid inefficient storage methods into your battery bank. I have also seen very simple electronics that can greatly reduce the surge power effect which can be very important in the conservation of your solar independent utility system.

Chapter 35) Construction

The construction of your solar system is very important as you will need access to it to clean the glass when it gets dirty, to remove the snow in the winter, and if you are trying to squeeze every watt out of your PV, to adjust the angle in accordance to the sun angle. I prefer the construction of a solar system in the yard mount system type for the ease of access. The wiring in your solar system when put together in the construction process needs to have as few split ends as possible connecting your PV to the battery bank. Any split ends needs to be soldered or anticorrosion cream applied to all junctions along with grounding your system for lightning strikes with a seamless copper wire which is extraordinarily important. That can also save you thousands of dollars.

The placement of your batteries in the construction of your solar system must be ventilated properly. The inverter needs to be located as closely as possible to the batteries to operate efficiently and properly. The closer the PV system is to your battery bank is important, as well to lower power loss. Wiring should be properly sized in accordance with the wattage it will be carrying along with the distance to avoid unnecessary power loss. Most all of this can be done by yourself if you are fairly mechanically inclined. The final connection to the electrical box however, should be done by a licensed electrician and observed by an electric company official when installing a grid-tie photovoltaic system.

Other thoughts in construction that need to be determined by your financial ability are the type of materials in your walls, floor, and roof that have the ability to store heat in them. The amount of insulation used, how many windows you choose to have, how air tight your home is, and also how much fresh air is introduced will greatly affect your homes efficiency.

A lot of this depends on your financial ability and the materials available depending on where you live in the world. I prefer to use materials that resist rot, decay, that have the ability to hold heat, hold cold, and can withstand a very substantial storm. Block and brick are my choice with adequate insulation. All of the choices that you make will greatly affect what your solar independent utility system will need to achieve.

Chapter 36) Powering Electric Vehicles from your Solar System

This is my chapter to the world of a solution to the beginning of the end of fossil fuel pollution from oil. There is no simpler solution that I have seen in my life that is just as easy as plugging up your cell phone to your battery bank that is charged by solar panels. This is absolutely not complicated at all. Instead of going to the gas station, filling up with pollution causing gases that pollute the environment causing numerous health problems, and contributing to climate change, we need to plug our cars into the power of the sun to give us free electricity.

I see no more elementary solutions to this crisis than having an automobile that you park at home, at the office, or any job then plugging it into the power of the sun. This is free, renewable, and 100% totally green, environmentally friendly endless supply of power that will be there forever as we know it on earth. I'm sure that the distance of the cars that may be traveled will go from a range of 40 miles as we have already seen to 120 miles as we hear is coming, to distances that will span across America in the foreseeable future. This will end our addiction to oil that we have so unwisely chosen.

Chapter 37) Alternative Fuels For Vehicles

Inside the slow conversions of the entire transportation fleet of planes, trains, semi-trucks, trucks, cars, and motorcycles in America for example; although not green this is still cleaner than gasoline, diesel, or coal. We have a seemingly endless source of natural gas all over in America that with a few conversions could power every transportation vehicle that is out there today.

I have driven across America and back twice off of the power of propane without spending any significant money to do so. This is very possible right here, right now, to end the giant sucking sound of America's finances to hostile countries which are polluting our environment and causing climate change. I see huge potential for powering all the vehicles in the fleet of transportation with hydrogen fuel which would also not be a significant change over cost. With much improved battery capacity it seems totally possible and achievable to power most of the vehicles by charging long lasting batteries from the power of the sun in the future. This will be a 100% clean and independent source of solar electricity that will go a long way to stabilizing the climate change. This will also clean up our air, as well as give us a secure independent electrical source that is renewable for countless years to come. Edison talked about (in chapter 44), an awesome example of how to do so with electric vehicles. This will be an important part in your solar independent utility system, as well as cleaning up the world.

Chapter 38) Water PSI

In the mindset of independent utility systems I would like to educate you on your most vital life sustaining substance, water. Most do not understand how the water makes its way to your faucet. This starts with electricity pumping it from the ground or water reserve. The water furthers its path towards your home, office, or business by electric pumps. Electric pumps take it through filters and a chlorination process. Finally it is pumped again by electricity into water towers that dot the landscape of our planet. Once in these towers it is pressurized by gravity pulling it towards the ground through pipes leading to your place of use.

There are regulators along the way that restrict the psi to 60 psi as it enters your place of use. When you turn on your faucet it is essentially gravity that makes the water flow out of your faucet. Many might think when my electric goes out and my water keeps on flowing it always will. I am informing you that it won't for very long. Once your local tower supply is dry because, of the electricity in the area is out for pumping the water, that tower of water you are so dependent on will eventually run out, so will your faucet. This is why having your independent water system is vitally important to you. A source of water can be used at your location such as rain water, spring water, or well water. This way you are depending on yourself for all of your water needs. This way you are not dependent on water companies that are sometimes impure, inflationary, unreliable, and also impose involuntary water restrictions at their will. They can also shut you off at any time for non-payment.

Water is so important to your survival; I believe one must strongly consider having their own source at the least to back up the utility that they are dependent on. The amount of psi you use will totally affect how much water that you consume. I have made my system to operate at 40 psi instead of 60 psi that the water company uses. This in itself saves half of the water that most people use.

Chapter 39) The Benefits of Going Green

There are so many benefits of going green that I hope I do not overlook the most important ones. I feel strongly that I must talk about it some to help motivate anyone on the fence to go green. My first thought is the future of our planet. I can't help but, to be concerned about what the kids in the future will inherit. The trash is making its way into the world's waters everywhere. The pollution that is filling the air is awful. The mercury that is raining down into our rivers, lakes, and streams from burning all of the coal to produce electricity is also getting into our food supply. The benefit of going green will help you, everyone around you, and all who will inherit the planet. Stopping the air pollution from using dirty coal burning power plants will greatly help to halt the affects of climate change. Going green would help stop air pollution from cars and trucks by using cleaner fuels (like solar electricity) would be a huge help to halting the climate change, as well as being uplifting to humanity.

Changing our trash consumption will also contribute with many other people will greatly impact the trash that is ending up in our oceans. Simply stopping the exhaust from cars alone would make the difference in people's health that breathes in the dirty air daily. Going green would give us healthier communities to live in worldwide. This would also lower energy in water cost for everyone. Going green would also give us a stronger environment and more sustainable cities for the future. Having a more stable environment I would believe that to be enough. Reducing the greenhouse gases should be everyone's goal to help make the planet a more livable place. This past year I experienced a drought that has not been seen since the 1800's. The drought almost totally dried up all of my communities' lakes, which also had a huge impact to the local crops. I believe it's wasteful not to utilize the power of the sun that is totally free and 100% green.

Chapter 40) Power Outages (Grid)

Power outages on the national grid although are rare, do happen, and affect many hundreds of thousands of people at a time. I find it interesting when doing a search about this they recommend electric generators and kerosene heaters. The national grid is set up like a line of dominoes, if one substation fails it overloads others until the system is stabilized. The problem lies in the wires. They go largely unprotected, are vulnerable to solar flares, high winds, as well as things like trees that took out 1/3 of America's power for nine seconds and left some in the dark for two months. When there is a wide spread outage, it affects the power plants themselves, as they have to shut down when the grid goes down because, they can't operate without power. What must happen in these cases is one plant is restored at a time to power up another plant so many hundreds of times until they are all back online again. This process conceivably if all power plants were down at the same time I have heard would take up to five years to get all of the power plants back online powering the grid this means the power that operates your home, office, or industry could be without power up to 5 years.

Any one day you can find stories like this: "Power back in areas where grid substation caused a power failure." When you are the one left in the dark it will be much more than the dark that will be bothering you. I have heard countless stories and they are all basically the same, "We often have problems." Being 100% solar independent from all utilities is not for just the rich, it is for everyone who wants to be dependent on yourself and not the utility companies.

Chapter 41) Tesla Technology

Tesla had a really unique idea that electricity could be transferred without wires through the air that could actually light up light bulbs. Tesla also had a unique ability of understanding DC current so much that he invented AC current by his knowledge of the frequency waves of both AC and DC. Tesla was the one that had the idea to take the power of Niagara Falls from all of the water flowing to turn great big wheels that would later become the hydro power producing AC power that is still powering New York City today, when he was just a very young boy.

Tesla was substantially a self-taught man. When Tesla saw the first DC motor he saw some flaws in it because, of it sparking then he quickly came up with the idea of an AC motor. The Tesla electric company capitalized with a half of a million dollars and it was opened for business in 1887. This consisted of a single phase, 2 phases, and 3 phases of alternating currents. Tesla established the 60 megahertz cycle for alternating currents which is still used today. What has been calculated and well known is that the sun puts off 64 million watts per square meter of the sun's surface.

Tesla found the need for insulating high voltage equipment by immersing it in oil. This is a method that is still very much used today as it is used in a lot of the transformers outside of all houses and businesses all across the world. Tesla was fascinated as a child, about the relationship between lightning and rain. Tesla goes on to say controlling lightning, he concluded, would be the most convenient way of harvesting the power of the sun. Tesla envisioned that it should be easy to perfect the idea that batteries that can store a whole year's supply of electricity. Tesla declared it would be, "a great deal less artificial than for men to dwell down into the bowels of the earth at so much trouble and loss of life in order to get a few handfuls of coal…"

The concept that Tesla had of electricity passing on the outside of the body without hurting oneself at all is still used even today in the repair and maintenance of the highest powered transmission lines that go across all of America making a huge electrical connection that we call the grid. Tesla said, "The earth was found to be literally alive with electric vibrations."

One of Tesla's experiments, Tesla rushed to the telephone and called the Colorado Springs electric company. He began remonstrating and pleading. The electric company had cut off his power he charged and wanted it restored at once. The reply from the powerhouse was curt and to the point. "You've knocked our generator off the line, and she's now on fire." Tesla had overloaded the dynamo. The town of Colorado Springs was in darkness.

Tesla was thoughtful of the conservation of non renewable resources was of critical importance to the world. He went on with his thoughts to express that he was convinced that wind and solar power should be developed. A past author wrote Tesla was the electrician of greatest promise. He was sued for non-payment of electricity from the city power company of Colorado Springs. He never saw smoke escaping from a stack that he wasn't offended by the waste of fuel that used up infinite resources. Tesla predicted that solar heat would partially supply the needs in our homes. He also conceived of a plan for transmitting energy in large amounts from one planet to another. He also had plans for extracting electricity from sea water and another from geothermal steam plants.

Works Cited

Cheney, Margaret. *TESLA MAN OUT OF TIME*. First Touchstone Edition 2001 ed. New York: Simon&Schuster, 1981. Print.

Chapter 42 Franklin

Franklin proved by flying a kite that lightning was electricity, and he invented a rod to tame it. He lived by four rules and they were: 1. It is necessary for me to be extremely frugal for some time, till I have paid for what I owe. 2. To endeavor to speak the truth in every instance; to give nobody no expectation that are not likely to be answered but, aim at sincerity in every word and action- the most amiable excellence in a rational being. 3. To apply myself industriously to whatever business I take in hand, and not divert my mind from my business by any foolish project of suddenly growing rich; for industry and patience are the surest means of plenty. 4. I resolve to speak ill of no man what so ever.

More of Franklin's guidelines that guided him: Silence: Speak not but what may benefit others or yourself; avoid trifling conversation. Order: Let all your things have their places; let each part of your business have its time. Resolution: Resolve to perform what you ought; perform without fail what you resolve. Frugality: make no expense but to do good to others or yourself; (i.e., waste nothing). Industry: Loose no time; be always employed in something useful; cut off all unnecessary actions. Sincerity: Use no harmful deceit; think innocently and justly, if you speak, speak accordingly. Justice: Wrong none by doing injuries, or omitting the benefits that are your duty. Moderation: Avoid extremes; forebear resenting injuries so much as you think they deserve. Cleanliness: Tolerate no un-cleanliness in body, cloths, or habitation. Tranquility: Be not disturbed at trifles, or at accidents common or avoidable. Chastity: Rarely use venery but for health or offspring, never to the dullness, weakness, or the injury of your own or another's peace or reproduction.

In Franklin's quest to test his idea about lightning he used a sharp wire that protruded from its top and a key was attached near the base of the wet string, so that a wire could be brought near it in effort to draw sparks. Suddenly he saw some of the strands of the string stiffen. Putting his knuckle to the key, he was able to draw sparks (and notably, to survive).

He wrote a paper in the edition of poor Richards Almanac, with an account of, "how to secure houses, ECT, from lightning." He also came up with the distinction between insulators and conductors; this idea of electrical grounding and the concepts of capacitors along with batteries. No one should underestimate the practical significance of proving that lightning, once a deadly mystery, was a form of electricity that could be tamed.

His thirst for knowledge had made him the best self taught scientist of his time. Franklin had studied how liquids produced different refrigeration effects based on how quickly they evaporate. Franklin wrote, "From this experiment one might see the possibility of freezing a man to death on a warm summer's day." His study of heat and refrigeration, though not as seminal as his work on electricity, continued through-out his life.

Works Cited

Isaacson, Walter. *Benjamin Franklin AN AMERICAN LIFE*. New York: Simon&Schuster, 2003. Print.

Chapter 43) Einstein

Einstein with his uncle Jacob went into business together setting up a small electro technical factory. He knew that his true interests lay not in mathematics but, in physics. There are strong reasons to believe that it was Einstein's rare mastery of Maxwell's electromagnetic theory that ultimately prompted Haller to offer Einstein a professional job in the Swiss patent office. What was noteworthy was that the young man advocated a wave theory of light without waiting to find wave explanations of all known optical effects.

Pulls and thrusts of motions give rise to the electromagnetic waves whose frequencies, or rates of oscillators, within quite a narrow range. Einstein was fully aware of the tremendous triumphs of the electromagnetic wave theory of light. Above all he applied his idea to the ejection of electrons from metals by light, a phenomenon called the photoelectric effect. This was otherwise known as solar power, or photovoltaics.

Einstein was able to deduce an utterly simple photoelectric formula. His photoelectric results went far beyond what was known experimentally at the time. "Let us round out this chapter by looking beyond 1905", this was an Einstein quote that shows his vision of our future. Einstein won the Nobel Prize in 1921, and the only work that was specifically mentioned was for the discovery of the law of the photoelectric effect. He gives formulas for the motion of electrons in an electromagnetic field, taking into account of the realistic increases in their masses as their speeds increase relative to the observer.

Works Cited

Kyle William Loshure

Dukas, Helen, and Banesh Hoffmann. *Albert Einstein Creator and Rebel.* New York: Viking,Inc., 1972. Print.

Chapter 44) Thomas Edison

Edison's quest is for inventions to transform the middle class American life. He built two rudimentary machines to generate electricity, one by friction, the other by magnetic action. He laid great emphasis on construction of simple experiments, demonstrating in detail a variety of batteries and electric toys. Edison understood in the summer of 1867 the principle of electromagnetic induction, demonstrating that an electric current could be produced by thrusting a magnet into a coil of wire and withdrawing it, thus giving birth to the dynamo along with the transformer. He broadened the domain and identity of electricity. He declared that all electricity was constant in the universe and "identical in nature," as were by extension of all the forces, "by which we know matter."

Edison discovered a varying electromagnetic current was used for both speaking and listening. He knew that there had to be a way to make the sound better. Orton's attitude after spending time with Edison said he was "an ingenious electrician." He wrote "the cornerstone of magic is an intimate practical knowledge of magnetism and electricity, their qualities, correlations, and potencies." The main subject of conversation among the trio of his friends was the exploited landscapes and the transmission of electric power.

Edison's work on the electric light, hardly Archimedes "eureka!" by any stretch of the imagination is best characterized by the ingeniously aphoristic statement, "I have the result but, I do not yet know how to get it." He went on to know that "to own, manufacture, operate, and license the use of various apparatus used in producing light, heat all powered by electricity was powerful to stockholders. Back in Edison's day when he conceived that, "wires would be laid underground in precisely the same manner as gas lines; the whole thing would come in at 5% the cost of gas." This was precisely "his method," exploited through the strategic perception of grass roots and corporate needs for convenient access to clean environmental energy. Edison was studying affects in his laboratory experimenting with varieties of electromechanical decomposition in lead acid battery cells and noting with relish that sulfuric acid that smelled. Edison worked on the perfection of "air telegraphy" between two moving trains using power generated from wet-cell batteries imitating dots and dashes that picked up as high as 600 per second.

Edison said, "This new war free crime free stain of mankind has perfected sun engines to harness solar energy." Tesla's previously written chapter talked about (refer to chapter 41) prototype, the current alternated indirection as the dynamo amateur revolved. This method of generating electricity was far more effective for electric power transmission over long distances, measured in hundreds of miles as opposed to Edison's direct current, whose range was limited to the length of city blocks. Edison remained adamant and entrenched opposition to high-tension electric current as "dangerous to life." Edison skeptically condemned use of "high-pressure" alternating current as a thinly veiled and cross attempt to save money in the manufacturing of conducting wire. Edison even mounted an ultimately fruitless lobbying effort through his affiliated companies nationwide to convince state legislators and local municipalities out of concern for "protection of the public" to outlaw high voltages.

An Edison quote that lives today, "what American (now living) will be the most honored in the 1990's. All gasoline motors which we have seen, belch forth from their exhaust pipe a continuous stream of partially unconsumed hydrocarbons in the form of a thick smoke with a highly noxious odor. The solution: - Electrically driven trucks, covering one half of the street area, having twice the speed, with two or three times the carrying capacity." He talked about "a new type of battery: The negative Pole or positive element was iron. The positive Pole or negative Pole was a superoxide of nickel. The electrolyte was potash, in aqueous solution containing ideally 20% potassium hydroxide. Unlike the lead battery, Edison's cell could be charged very quickly with no physical damage. He believed his new storage battery the most valuable of all his creations and believes it will revolutionize the whole system of transportation. Edison truly believed the future belonged to the electric car!" This was what was said in the year 1901. Edison acquired several different types of electric cars so that he could experiment with various types of batteries. The electrolyte was even further purified by adding a small amount of lithium hydrate to the potash solution. A last quote I read that says a lot about humanities current situation "It would be sad if we have to exhaust all of our fossil fuels before we realize the full potential of solar energy."

Works Cited

Baldwin, Neil. *EDISON INVENTING THE CENTURY.* New York: VosBurgh's Orchestration Service, 1995. Print.

Chapter 45) Watts

Watts are the main focus that you need to pay attention to when sizing your photovoltaic system. You simply need to go around your home or office and start adding them up. You can also go by your electric bill and view the current kilowatts. Add that up per year, times that by a thousand to get your yearly watts, and divide that by 365 to get your watts used per day. This is a base line to establish how much wattage you will need to supply from your solar system producing solar power. One watt is equal to one amp of current per second. A watt meter can be used to plug in your appliances into to see what the usage is. This will also help you to determine if you have an appliance that carries a phantom load, an appliance that uses watts even when turned off.

Wattage is equaled to voltage times current. Watts are a name of an electric flow. Another way of describing it is quantity of electrical flowing per second. Electrical power is measured in watts. The electrical system power is equal to the voltage multiplied by the current. Just to make it clear what we are talking about, another very simple way to know this equation is: Watts = amps times volts. A good understanding and knowledge of watts is important to understand for your solar independent utility system.

Chapter 46) Volts

Understanding volts is also important in your solar independent utility system. You will be taking 12 volts for example and converting it with an inverter to produce 120 volts (household electricity). Voltage is measured in volts. This sounds as simple as it is. This is very clear and easy to understand. The volt is defined as the value of the voltage across a conductor when a current of one ampere dissipates one watt of power in the conductor. The water analogy voltage is likening to the water pressure. Another way to figure it is: Amperes times ohms = volts. The unit of electromotive force, the volt measures how much "pressure" is in an electrical circuit. The higher the voltage, the more the current will flow into the circuit. Ordinary household outlets are usually rated at 115 volts, car batteries at 12 volts, and flashlight batteries at 1.5 volts. Common voltages within a computer are from 3 to 12 volts of direct current.

Chapter 47) Amps

Electric current is measured in amps. Also don't forget resistance is measured in ohms. Amps are vitally important to understand, if you overload the DC wire or an AC wire with amps a fire will result. The fuse will also blow if wired properly. Both the watts and the amps will determine how long your battery bank will last. When setting up your solar independent utility system you must know that the load (from amps and watts) does not exceed the wires capability in which you are sending power or supplying power to appliances as power losses will occur also potential fires may result. Amps should definitely be taken into consideration in sizing your solar system. Many motors such as washing machines or other motor driven appliances have surge amps for starting up the appliance as well as electronics such as microwaves. Televisions have surge power for start up as well. This needs to be taken into account as they will put a huge drain on your solar independent utility system.

Chapter 48) Megahertz

Hertz is defined as the number of cycles per second. This is very important as it is standard for electric companies to send power to homes and businesses as 60MHz. Megahertz is equal to 1,000 hertz. In order for your grid-tie solar system to send power onto the grid your inverter has to send the power just ever so slightly above the 60MHz to get the power flowing onto the grid or back to the electric company. You must understand this to grasp the way the power goes back to the electric company. The power does not literally go to the electric company but, your closest neighbor on the electric line. This gives you an electric credit. The inverter will monitor the megahertz from the electric company and if it goes down it will shut the power down as to keep lineman from being shocked. The number of oscillations of the perpendicular electric and magnetic fields per second are expressed in hertz. This is a scientific expression to understand how it is visualized. Sending power onto the grid is all about the megahertz. This will clarify how power is sent to the grid from your solar independent utility system with a grid-tie inverter.

Chapter 49) Sine Waves

The sine wave is a mathematical function that describes a smooth repetitive oscillation. The sine wave is very important because, it retains its wave shape when added to another sine wave of the same frequency, also arbitrary phase. Sine waves are the only periodic wave form that has this property. This property leads to its importance in Fourier analysis and makes it acoustically unique. This wave pattern often occurs in nature, including ocean waves, sound waves, and light waves. A rough sinusoidal pattern can be seen in plotting average daily temperatures for each day of the year. The sine wave has a pattern that repeats, simply put. Trigonometric identify is the mathematical way of writing this out. This is typically used by electrical engineers. Another way of putting this is the sine wave is a basic function employed in harmonic analysis. This will help you to better understand how the sine waves work in your solar independent utility system to make it possible for you to produce your own power with an inverter to make the sine waves that you need to operate your electrical devices.

Chapter 50) Modified Sine Waves

Modified sine waves are more useable than block (square) sine waves. They run most all electrical electronics but, not the most sensitive ones. Pure sine wave inverters run anything but, are very expensive. Modified sine wave inverters are an economical way to convert your solar power into useable 120 volt or 240 volt power. Modified sine wave inverters produce a power wave that is sufficient for most devices. The power wave is not exactly the same as electricity from the power grid. What does that mean to the everyday user? Not much. This runs most all electronics just fine. A humming might be heard in a ceiling fan for example or a small line might appear on your television that travels across the screen almost not noticeably. Most inverters of all types of sine waves have an auto shut off that beeps and disconnects from the battery bank when the battery bank gets too low. The reason for this is too low of a discharge will damage your batteries. This will protect your batteries.

A grid-tie inverter simply switches to the grid for power when this point is reached. With a parallel solar system the beep means it is time to switch your AC load to the grid while leaving the DC connected all of the time year after year. With a standalone system when the beep occurs telling you that your batteries are low you can simply stop using the AC power for the night from your inverter. A generator can be used to supply your AC power at this point and also can be used to charge your battery bank.

Some computers and stereo equipment use switching power that supplies that utilizes SCR's as well as Triac's (switches) as well. These pieces of equipment may experience the same troubles of overheating problems with modified sine waves. Knowing what exact models of equipment will have problems with the modified sine waves or square (block) wave forms is hard to predict. The only way to know for sure is to try it. I personally have not found any issues with the modified sine wave other than the hum in the ceiling fan or the small line that travels across the television screen.

I like the modified sine wave the best for its dependability and reasonable cost. The oscilloscope is a device that can show in graph form of what the output of the inverter looks like. There are also 24 volt modified sine wave inverters available that will not drain your battery bank as fast. The drawback is that you can't utilize the 12 volt power for many uses that you get no power loss by stepping up the power through an inverter that is typically 10% to 30% power loss. The modified sine wave inverters are very useful in powering your entire household or any business application. They are great for the cost and dependability. The modified sine wave inverter is the key component that takes your solar power and steps it up to useable grid-type power. This will be an amazing part of your solar independent utility system.

Chapter 51) Square (Block) Sine Waves

The square (block) sine wave is the lowest cost inverter. This is also good for running a limited amount of electronics. The square sine wave is a kind of non-sinusoidal wave form, most typically encountered in electronics and signal processing. The ideal square wave alternates regularly instantaneously between two levels. Square sine waves contain a wide range of harmonics.

These can generate electromagnetic radiation or pulses of current that interfere with other circuits, causing noise as well as errors. The issues with square sine waves are again humming and lines across the television screen. Overheating in the unit may occur. There are no good ways to know if the electronic device that you want to use will work with a square sine wave inverter but, since it is the most affordable it is worth a try. Modified sine waves I like the best for the value and the ability to run everything I have tried. The pure sine wave is a fail proof way to have no issues whatsoever. The cost of a pure sine wave is much higher than the modified sine wave or the square sine wave.

The square sine wave is just fine for very basic electronics like lights or simple devices like that. I would stay away from sensitive electronics such as computers or digital devices with a square sine wave to be sure no damage will result.

Chapter 52) Pure Sine Waves

Pure sine wave inverters are the most costly but, have the ability to run all electronics without any humming noise also any lines on the television screen. They don't overheat and will not harm any electronic device that you have. Pure sine wave inverters have the smoothest oscillation of the three types. Square or (block) sine wave being the choppiest sine wave. Modified sine waves are between the square sine wave and the pure sine wave. The pure sine wave is the highest quality inverter on the market. This is also a highly dependable inverter like all other inverters. Owning expensive electronics it would be worth it to invest in a pure sine wave inverter to insure the protection of all your electronics.

They come in high enough wattage to power your entire house and typically have the lowest power loss of all three types of inverters. They are available in 12 volt, 24 volt, or a 48 volt version. I like the 12 volt best as you can use so many things like your propane refrigerator that runs off of the 12 volt as well as water pumps without any conversion and power losses. With the case of unusually high power loads you might want to consider the 48 volt system as your battery bank will not drain as fast. Higher circuits such as 24 volt and 48 volt can be made with your solar power to charge the battery bank. Just keep in mind the larger the circuit the more shading affects the power that is being produced. Things like power lines along with trees will be a bigger issue in a 24 volt or a 48 volt system.

Grid-tie pure sine wave inverters are available as well. They will commonly be the most efficient by having the lowest power loss in converting your DC solar power to AC power (household power). This works by simply any power that your solar system is producing in excess of your demand (household use) will be sent back to the grid giving you an electric credit. The larger your PV system the larger your credit will be. The goal for your grid-tie pure sine wave inverter should be your sending more power to the grid than you are using in a year to offset your electric bill to be zero for the calendar year or a monthly credit that will give you a check from the electric company on a regular basis. Having a solar independent utility system with a proper inverter will give you a 100% utility independent system that will serve you for countless decades with perfect performance.

FAQ'S

1. How do you get energy from the solar panels to your home?

   Answer: The energy you get from your solar panels to your home travels through wires that connect your solar panels to your home or business.

2. How do you get your water filtered?

   Answer: You get your filtered water simply passing it through a two micron filter (for example) that is made of cotton or activated carbon that cleans the water.

3. How does a standalone system work?

   Answer: A standalone system works by gathering energy from the sun, gathering water from the rain (for example), and utilizing them in a way to supply your home as well as business with water and electricity.

4. How do you recycle water?

   Answer: You recycle water by using a water pump to take water from storage and pass it through a filter then send the water back into the storage (this is great for purifying water or keeping the water moving so it doesn't freeze).

5. How do you bring water from storage outside?

Answer: Bringing water in from outside storage (or collection barrels) is done by shutting you're inside water supply valve off, and turning on your outside water valve. Then open the fill valve for your inside water storage.

6. How does a parallel system work?

   Answer: A parallel system works by using both solar power as well as electricity from the grid next to another to supply your electric needs, but they never connect.

7. What is phantom power?

   Answer: Phantom power is appliances and transformers that use electricity even when there shut off (televisions are an example).

8. How can you have hot water that never runs out?

   Answer: Having hot water that never runs out is accomplished by using an on demand water heater (tank-less) that heats the water as it passes through it so it will never run out.

9. How does a grid-tie system work?

   Answer: A grid-tie system works by sending the solar power in excess of what you are using onto the grid giving you an electric credit. You can also have a battery backup system that saves power in the batteries to be used when the grid is down that supplies your electricity needs while you have sufficient power saved to run your demand.

10. How can your DC power never go out?

    Answer: Your DC power can never go out in a calendar year by having proper battery maintenance (checking for dead cells) sizing your PV array expected wattage with your expected energy usage with the proper storage capacity in the battery bank.

11. What is the difference between AC and DC power?

    Answer? The difference between AC and DC current is DC is a direct current (or power) as its constant, always on with no pulsations. AC current is a pulsating current that is powered by pulses of current (or power). Fact: This is why resistance is present with DC current and resistance is not as much of an issue with AC current.

12. How does a drain back water valve system work?

    Answer: The drain back water valve system works by a set of valves that close the water line supply as well as it opens up the pipes to let all the water out of the on demand heater to protect is from freezing while in storage or in a below freezing environment.

13. How do the hot air panels work?

    Answer? The hot air panel's work by letting the warm sunshine pass through a clear plastic onto a blackened surface with insulated backing creating a space of air that's heated and is brought into the place of use by a fan forced vent.

14. How does geothermal cooling work?

    Answer: Geothermal cooling works by recycling a fluid through pipes that bring the coolness of ground into a fan forced system (much like an air conditioner) that cools your home or business.

15. How do you store water?

    Answer: Water can be stored simply by putting into a closed degrade free container that keeps the bugs out and add the right amt of chlorine then it will be safe as well as not go stale.

16. What are some ways to save water?

    Answer: Some ways to save water are the low flow toilet, (number one and number two flush) lowering your PSI (water pressure), moving your water heater closer to your shower, also using a shower head from directly overhead to be able to slow the rate of flow. Using shut off valves at the end of the kitchen faucet, automatic shut off sensors, and flow reduction attachments.

17. What are some ways to save electricity?

    Answer: Some ways to save electricity are: unplug all phantom loads, change to compact florescent lights using the lowest possible wattage you feel comfortable with, and using led lights. Get rid of: electric furnaces, electric water heaters, electric stoves, electric driers, electric refrigerators and anything else that uses electricity to produce heat! Replace the old appliances with more efficient gas: stoves, driers, vent free radiant heaters, LP gas refrigerator, and a gas on-demand water heater. Install hot air producing solar panels.

18. What is a battery bank?

    Answer: A battery bank is several batteries wired together in series to produce 12, 24, 48, volt power, or wired in parallel to maintain the current listed on the battery. They hold all the power your solar panels produce during the day for night time use. The battery bank adds to the solar power supply while excess power (more than what your solar panels are producing) is being used from your household as well as business use.

19. How big of a system do I need to run my house or business?

    Answer: To determine what size of solar system you need to operate your house or business you must consider how much area you have for your PV system to gather sunshine (available wattage). You also need to seriously take into consideration making all the changes listed in this manual to lower your electricity usage. Once you see the available square footage you will know how many watts you are able to collect. The solar pathfinder will help you to calculate how many sun hours are available over the year. This along with your financial ability will greatly drive you towards what size system you can put in. Going by the guide lines in this manual is the lowest cost way of going totally solar power known today. To know exactly how big of a system to install you need to know, 1. The watts on a daily basis you will be using times a calendar year. 2. The amount of ambient light available to collect from. 3. The amount of output (power produced in the solar panel) as well as the area they take up. 4. Going by the solar pathfinder to calculate all hours of light this will give you a rough estimate of what to expect. You need to add an additional 30% to the estimated amount of PV to cover for future increases in electricity and prolonged poor collection conditions. This is the best way to know for sure how big of a system to install. With a properly sized PV system you will have a more dependable electric supply than you currently have.

20. How do I utilize spring water, rain water, or well water instead of using city water?

    Answer: You can utilize spring, well, or rain water simply with a collection system for all of them with a storage system to hold a water supply large enough to last you. Using a water pump the water can be pressurized with an on demand pump that keeps the water flowing like normal when you turn on the faucet. By using a water filter you can have more pure water than you get currently.

Glossary

A

A fixed PV system – A system that is mounted in place and does not track the sun. Adjustments as to the angle are made manually. A permanent PV system. A manually adjustable PV system.

Abundant – Present in great quantity; more than adequate; over sufficient. Well supplied; abounding. Richly supplied. Existing in plentiful supply.

AC motor – A motor that runs off of AC current.

Accordance – Agreement; conformity. The act of granting; bestowal: in accordance with the rules.

Achievable – To bring to a successful end; carry through; accomplish. To get or attain by effort; gain; obtain. To bring about an intended result; accomplish some purpose effort.

Acoustically – Pertaining to the sense of organs of hearing, to sound, or to the science of sound.

Activated carbon filters – A filter that takes out chemicals, taste, and odors.

Adamant – Utterly unyielding in attitude or opinion in spite of all appeals, urgings, ect. Unshakeable in purpose, determination.

Addition to oil – Our inabilities to find ways to do things without the use of oil.

Adequate insulation – The amount of insulation that is recommended for your particular location in the world.

Affiliated – Being in close formal or informal association.

Air pollution – Contamination of air by smoke along with harmful gases, mainly oxides of carbon, sulfur, and nitrogen.

Alterations – The act or process of altering; the state of being altered. A change; modification or adjustment.

Alternates – To reverse in direction or sign periodically. To interchange repeatedly and regularly with one another in a time, place; rotate.

Alternating current – An electric current that reverses direction at regular intervals.

Alternating current (single phase) – The normal power supply used in households.

Alternating current (two phase) – The power used in medium size electric loads such as light machinery.

Alternating current (three phase) – The power used in industrial locations such as heavy machinery or large demands such as an electrical furnaces that melt metal.

Amateur – A person inexperienced or unskilled in a particular activity.

Ambient light – The light surrounding an environment or subject, esp. in regard to photography along with other work like solar energy production.

Amperes – The base SI unit of electrical current, equivalent to one coulomb per second, formally defined to be constant of infinite length, of negligible circular cross section, and placed one meter apart in a vacuum, would produce between these conductors a force equal to 2 times 10 to the negative 7 Newton per meter of length. Abbreviation A, amp. The practical marks unit of electric current that is equivalent to a flow of one coulomb per second or to the steady current produced by one volt applied across a resistance of one ohm.

Amplified – To make larger, greater, or stronger.

Amps – A typical household's electrical supply includes a total of 120 to 200 amps; a typical house circuit carries 15 to 50 amps.

Analogy – A similarity between like features of two things, on which a comparison may be based. A form of reasoning in which a similarity between two or more things is inferred from a known similarity between them in other respects.

Analysis – A philosophical method of exhibiting complex concepts, propositions as compounds, along with functions of more basic ones. The process as a method of studying the nature of something or of determining its essential features and their relations. The separation of any material or abstract entity into its constituent elements.

Anticorrosion cream – A cream dressing that can be applied in a paste form onto wires that are connected to stop corrosion.

Aphoristic – Of, like, or containing aphorisms. Given to making or quoting aphorisms. Tending to write or speak in aphorisms.

Apparatus – Any system or systematic organization of activities, functions, processes, ect. , directed toward a specific goal. Any complex instrument or mechanism for a particular purpose.

Acquired – To come into possession or ownership of; get as one's own. To gain one's actions or efforts.

Area – Any particular space or surface; part. Extent, range, scope. A part or section, as of a building, town, ect. , having some specified function or characteristic. An unoccupied or unused flat piece of open ground.

Artificial – Lacking naturalness or spontaneity; forced; contrived; feigned. Imitation; simulated; sham. Made in imitation of a natural product, esp. as a substitute.

Attitude – Manner, disposition, feeling, or position, ect. , with regard to a person along with a thing; tendency or orientation, esp. mind. Position of posture of the body to or expressive of an action, emotion, ect.

Auto shut off – A switch that shuts off automatically when the power gets too low. A switch that shuts off the water pump when the water pressure reaches the rated psi.

Automatic faucets – Faucets that turn on when you put your hands under for washing and shut off when you remove your hands.

Automatic setback thermostats – A thermostat that can be programmed to either turn up the temperature or turn down the temperature as you program it to do so to save money on heating also cooling.

Awful – Extremely bad; unpleasant; ugly. Inspiring fear dreadful; terrible. Extremely dangerous, risky, injurious, ect. Very bad; unpleasant. Not standard.

B

Backup generator – A power source that can be used to supply power whenever your solar systems power production is low or when your battery bank is discharged. A device to be used when the grid is down to supply your electrical need with power from a fossil fuel.

Battery acid tester (battery tester) – A floating ball tool that measures the density of the acid to determine if a cell in the battery is bad.

Battery storage or (battery bank) – A large number of batteries that are wired in parallel or series to provide you with stored electricity.

Benefits – Something that is advantageous or good; an advantage. A payment or gift, as one made to help someone. An act of kindness.

Broadened – To become or make broad. Synonyms- extended, expand, enlarge, widen; enlightened, informed, educate; sophisticate.

Byproduct – A secondary or incidental product, as in a process such as burning coal. The result, along with another action, often unforeseen, or unintended. Something produced in the process of something else such as mercury released into the air as the result of burning coal in the production of making electricity at the electric power plants.

C

Capacity – The ability to receive or contain. The maximum amount, number that can be received, or contained. Capacitance. A measurement of the electrical output, also a piece of apparatus such as a motor, generator, or accumulator.

Capitalized – To take advantage of; turn to one's advantage: to capitalize on one's opportunities.

Carbon neutral – Emitting no carbon dioxide into the atmosphere; also, employing a technique to absorb carbon dioxide so it is not emitted; also written carbon-neutral, carbonneutral.

Charge controller – A device that monitors your battery bank and shuts off the power from your PV system to stop the charging when your batteries reach 100% full. A regulator for your solar system that controls the charge. A protective device that stops the battery bank from overcharging.

Charged – Pertaining to a particle, body, along with a system possessing a net amount of positive or negative electric charge. The process of charging a battery to capacity.

Chlorination – To combine or treat with chlorine. To introduce chlorine atoms into an organic compound by an addition or substitution reaction. To disinfect (water) by means of chlorine.

Choppiest – Forming short, irregular, broken waves. Shifting, changing suddenly, or irregularly; variable. Uneven in style, quality, or characterized by poor related movements.

Cleaner fuels – A source of fuel that pollutes the environment less than the source of fuel that you are currently using.

Coal – Anthracite bituminous coal lignite; also peat a combustible compact black or dark brown carbonaceous rock formed from compaction of layers of partially decomposed vegetation: a fuel, source of coke, coal gas, and coal tar. A dark-brown to black solid substance formed from the compaction, along with hardening of fossilized plant parts in the presence of water, and in the absence of air. Carbonaceous material accounts for more than 50% of coal's weight and more than 70% of its volume. Coal is widely used as fuel. Coal also contains toxic air pollution mercury that is released into the air in the burning process.

Coil of wire – An electromagnetic coil is formed when a conductor is wound around a core, also to create an inductor, or electromagnetic.

Cold cranking amps – The amount of amps that a battery has for use under extremely cold conditions.

Communities – A locality inhabited by such a group.

Compact florescent bulbs – A high energy efficient light bulb that only uses a fraction of the power that an incandescent light bulb does.

Conceivably – Capable of being conceived; Imaginable.

Conceived – To form a notion or idea of; imagine. To apprehend mentally; to understand.

Concerned – Interested or affected. Troubled or anxious. Having a connection or involvement; participating.

Condemned – To declare incurable. To express an unfavorable or adverse judgment on; indicate strong disapproval of; censure.

Condensation – The act of condensing of for example water droplets on a cold window.

Conductor – A substance, body, or device that readily conducts heat, electricity, sound, ect.

Connection – Anything that connects, connecting part, or link.

Correlations – Mutual relation of two or more things.

Conservation – The act of conserving; Prevention of decay, waste, or loss.

Construct – To build or form by putting together parts.

Construction – The process, or act of constructing, or manner in which a thing is constructed.

Consume – To use up; to spend wastefully. To destroy, as by decomposition or burning.

Consumption – The act of consuming, as by use, decay, or destruction. The act or process of using up something.

Contribute – To give to a common supply. Contribute to, to be an important factor in; help the cause.

Contributing – To give to a common supply, fund, ect. , as for charitable purposes. To give for a common purpose or fund.

Conversions – Change of something to another thing.

Converting – To change from one form to another; transform.

Convinced – Obsolete to overcome, confute, or convict; to overcome the doubts of.

Cost effective – Producing good results for the amount of money spent; efficient or economical.

Contaminated (contamination) – To make impure, infected, corrupt, radioactive, ect. By contact with or addition of something; pollute; defile; sully; taint.

Credit – Belief or trust; confidence; faith; rare the quality of being creditable or trustworthy.

Critical – Tending to find fault; censorious; characterized by careful analysis and judgment.

Current – Electric current. The time rate of flow of electric charge, in the direction that a positive moving charge would take and having magnitude equal to the quantity of charge per unit time: measured in amperes.

D

DC current – An electric current flowing in one direction.

DC lighting – Lighting that runs of DC current, example a 12vold light bulb, a 1watt led light, or diode.

DC motor – A motor that runs off DC current for its power supply.

Decades – Any ten year period.

Decomposition – To break up or separate into separate components. The act or process of decomposing. The state of being decomposed; decay.

Deduce – To infer by logical reasoning; reason out or conclude from.

Deep cycle batteries – Batteries that are tolerant of deep discharge.

Demonstrating – To show by reasoning; prove; to explain or make clear using examples, experiments.

Dependability – Trustworthy reliable. Capable of being depended on; worthy of trust; reliable.

Dependable system – A system that you can depend upon without fail.

Dependent – Contingent; relying on for support or aid.

Describing – To tell or write something about; give a detailed account of; to picture in words.

Deterioration – To make or become worse over time.

Determined – Having one's mind made up; decided; resolved; resolute; unwavering.

Direct vent – Nonstop one way out without curves; going straight up or out.

Discharge – To rid (a battery, capacitor, ect,) of a charge of electricity.

Discharge rate – The rate of the removal or transference of an electric charge, as by the conversion of chemical energy to electrical energy.

Discharged battery bank – Ridding several batteries of their electrical charge by the equalization of a difference of potential, as between two terminals.

Disconnects – To interrupt the connection between; detached.

Discovery – To find, uncovering, breakthrough; the act or an instance of discovering.

Dominoes – You have a row of dominoes set up; you can knock over them all with one as in the game of dominoes.

Drain back water valve system – A system of valves that one turns to let water out of the water heater and shutting off the water supply at the same time.

Drought – A prolonged period of scanty rainfall.

Dual metering system – An electrical metering system that measures the amount of power you are producing with your PV and the amount of electricity you are using from the electric company.

Dynamo – An electric generator, esp. for direct current.

E

Educate – To instruct, school, drill, indoctrinate; to provide schooling or training for.

Ejection – To drive or force out; expel or emit. The act of expulsion, forcing out, projection, exclusion, riddance.

Electric pumps – A pump that is operated by electricity; 12DC or 120volt AC current.

Electric vehicles – An electric vehicle, also referred to as an electric drive vehicle, uses one or more electric motors for propulsion. Electric vehicles include electric cars, electric trains.

Electrical box – A box in which all the circuit breakers or fuses are located in a DC, also AC system.

Electrical circuit – An electrical circuit is a path which electrons from a voltage or current source flow. Electric current flows in a closed path called an electric circuit.

Electrical engineers – A field of engineering that generally deals with the study along with the application of electricity, electronics, and electromagnetism.

Electrical source – A place were electricity comes from.

Electricity – Electric charge, electric current; the science dealing with electric charges and currents. A phenomenon associated with stationary, moving electrons, ions, or other charged particles.

Electrolyte – Any of certain inorganic compounds, mainly sodium, potassium magnesium, calcium, chloride, and bicarbonate, that dissociate in biological fluids into ions capable of conducting electrical currents, along with constituting a major force in controlling fluid balance within the body. Also called an electrolytic conductor. A conducting medium in which the flow of current is accomplished by the movement of matter in the form of ions.

Electromagnetic – Of or pertaining to electromagnetism, along with electromagnetic fields. Of containing, or operated by an electromagnetism: An electromagnetic pump.

Electromagnetic field – The coupled electric, along with magnetic fields that are generated by time- varying currents and accelerated charges. A field of force associated with a moving electric field and a magnetic field at right angles to each other to the direction of propagation.

Electromotive force – The fundamental force associated with electric and magnetic fields. The electromagnetic force is carried by the photon, along with being is responsible for atomic structure, chemical reactions, the attractive also repulsive forces associated with electrical charge also magnetism, and all other electromagnetic phenomena. Like gravity, the electromagnetic force has an infinite range and obeys the inverse-square law.

Electronics – The science dealing with the development, along with the application of devices, gaseous media, and in semiconductors.

Electrons – Also called negatron. An elementary partial that is a fundamental constituent of matter, having a negative charge. A unit of charge equal to the charge of one electron.

Elementary solution – A solution that is very basic in nature easy to understand.

Emphasis – An object, idea, ect, that is given special importance of significance.

Entrenched – To place in a position of strength; established firmly or solidly: safely entrenched behind undeniable facts.

Environmentally friendly – Nature-friendly

Envisioned – To picture mentally, esp. some future event or events: To envision a bright future.

Essentially – Absolutely necessary; indispensable. Pertaining to or containing an essence.

Establish (established) – To make secure or permanent in a certain place. To prove correct free of doubt. To be widely or permanently accepted.

Eventually – Finally; ultimately; at some later time.

Everywhere – In every place or part; in all places.

Exhaust – The escape of steam or gases from the cylinder of an engine.

Expectancy – The quality or state of expecting; expectation; anticipatory belief or desire.

Experienced – Wise or skillful in a particular field through experience. Endured; undergone; suffered through; taught by expirence.

Experimentally – Relating to or based on an experiment. Given to experimenting. Of the nature of an experiment; constituting or undergoing a test: founded on experience; empirical.

Experiments – A test under controlled conditions that is made to demonstrate a known truth, examine the validity of a hypothesis, or determine the efficacy of something previously untried.

Exploited – A person taken advantage of; after going out of their way to help another out.

Extracting – To obtain despite resistance. To remove for separate consideration or publication.

Extraordinarily – Beyond what is ordinary or usual. Highly exceptional; remarkable.

F

Fascinated – To hold an intense interest or attraction for. To hold motionless; spellbind.

Faucet – A regulator for controlling the flow of a liquid from a reservoir such as a sink faucet.

Financial ability – The ability or means to afford something.

Florescent bulbs – A light bulb that runs off a gas instead of an incandescent bulb; is much more efficient because it does not produce the heat that is wasted energy.

Formula – A statement, especially an equation, of a fact, rule, principle, or other logical relation.

Foreseeable – To see or know beforehand.

Fossil fuel – A hydrocarbon deposit, such as petroleum, coal, along with natural gas, derived from the accumulated remains of ancient plants and animals which are used as fuel. Carbon dioxide and other greenhouse gases generated by burning fossil fuels are considered to be one of the principal causes of climate change. Any naturally occurring carbon or hydrocarbon fuel, such as coal, petroleum, peat, along with natural gas, formed by the decomposition of prehistoric organisms.

Frequencies – The number of times a specified periodic phenomenon occurs within a specified interval. The number of repetitions of a complete sequence of values of a periodic function per unit variation of an independent variable.

Frequency waves – The number of items occurring in a given category. The number of times a value recurs in a unit change of the independent variable of a given function.

Fundamental – A basic principle, rule, law, or the like, that serves as the groundwork of a system; essential part. Being the original or primary source.

Furthers – To help forward; promote; advance. At or to a more advanced point; to greater extend.

Fuses – A protective device, used in an electric circuit, containing a conductor that melts under heat produced by an excess current, thereby opening the circuit. To overload an electric circuit so as to burn out a fuse. Compare circuit breaker.

Future – Time that is to be or come hereafter. Something that will exist or happen in time to come.

G

Gas generator – A machine that converts gas form of energy into another, esp. mechanical energy into electrical energy such as a dynamo. Any device for converting mechanical energy into electrical energy by electromagnetic induction.

Gaseous fumes – The toxic air emitted from the exhaust pipe from cars, trucks, trains. Composed of molecules in a constant random motion. It has no fixed volume and will expand to fill the space available.

Generate – To produce, esp. in a power station; to produce by a chemical process. To produce or bring into being; create.

Generators – A machine that converts mechanical energy into electricity to serve as a power source for other machines. Electrical generators found in power plants use water turbines, combustion engines, windmills, or other sources like coal, along with nuclear for mechanical energy to spin wire coils in strong magnetic fields, inducing an electric potential in the coils.

Geothermal cooling – The way to use the 55 Deg. Constant Temp. of the ground to cool your home or business. It is a method that takes liquid in large quantities and passes it through pipes into the ground recycling the liquid constantly re-cooling it to cool you place of use.

Geothermal loop – This is the pipe that is very long and coils deep in the ground at the 55 Deg. Constant Temp. Line in the earth. Its length is based on you capacity needs.

Geothermal steam plants – Is power generated from natural steam, hot water, hot rocks, or lava in the earth's crust. In general geothermal power is produced by pumping water into cracks in the earth's crust and then conveying the heated water, steam back to the surface so that its heat can be extracted through a heat exchanger, also its pressure can be used to drive turbines that make electricity.

Grasp – Mental hold or capacity; power to understand.

Gravity – The force of attraction by which terrestrial bodies tend to fall toward the center of the earth. Serious or critical nature.

Green technology – Any technology that is environmentally friendlier than a comparable existing technology; Solar power is an example of clean technology.

Greenhouse gases – Any of the gases whose absorption of solar radiation is responsible for the greenhouse effect, including carbon dioxide, methane, ozone, and the fluorocarbons. The elevated levels especially of carbon dioxide and methane that have been observed in recent decades are directly related, at least in part, to human activities such as the burning of fossil fuels along with the deforestation of tropical forest.

Grid-tie (with battery backup) – A grid-tied electrical system, also called tied to the grid or grid tie system, is a semi-autonomous electrical generation along with a grid storage system which links to the mains electrical grid. When insufficient electricity is generated, or the batteries are not fully charged, electricity drawn from the mains grid can make up any shortfall. When the batteries are fully charged such as in the peak production all excess electricity is sent directly on the mains to give you an electricity credit if more power is produced than you consume over the course of the month.

Grid-tie inverter (tied) – A grid-tie inverter is a special type of inverter that converts direct current (DC) electricity into alternating current (AC) electricity and feeds it into an existing electrical grid. GTI's are often used to convert (DC) produced by renewable energy such as solar panels. They may also be called synchronous inverters. Grid-interactive inverters typically cannot be used in a standalone applications were utility power is not available.

Grid-tie only (without battery backup) – It is a photovoltaic (PV) system interacting with the utility, and can be without batteries, that utilizes a relatively new breed of inverter that can sell any excess power produced directly back to the grid.

Grid-tie photovoltaic system – A grid tied photovoltaic system is a system consisting of electric solar panels, a grid-tie inverter, batteries or no batteries that connects directly to the grid which then sends power in excess of charging the batteries, along with the power being consumed also the more efficient without battery system that also sends all excess power above what your consuming directly onto the grid, (note: w/out batteries it does not work when the grid is down even if the sun is out , if you want that it requires the battery backup which is much less efficient at sending power onto the grid).

Grid-tie system – Grid-tie systems are permitted in many countries to sell their energy to the utility grid through a policy known as net metering. So for example, if during a given month a power system feeds 500 kilowatt-hours into the grid and uses 100 kilowatt-hours from the grid, it would receive compensation for 400 kilowatt-hours. In the US, net metering policies vary by jurisdiction. In the US grid-interactive power systems are covered by specific provisions in the national electric code, which also mandates certain requirements for grid-interactive inverters.

Grounding – A conducting connection between an electric circuit, equipment, the earth, or some other conducting body. Grounds are used to establish a common zero-voltage reference for devices in order to prevent potentially dangerous voltages from arising between them and the other objects.

Grounding rod – A copper rod 6 feet long typically that is driven into the ground to prevent lightning strikes to a solar system.

Grounding wire – A continuous copper wire typically that connects to the frame of every solar panel then connecting directly to the grounding rod without any splits in the wire.

H

Halting – To stop; cease moving, operating, ect. either permanently or temporarily.

Harmonics – Of pertaining to, noting a series of oscillations in which each oscillation has a frequency that is integral multiple of the same basic frequency. A noise that can be heard in a modified sine wave inverter in things like ceiling fans or appear in things like televisions.

Heat collector – A solar thermal collector is a solar collector designed to collect heat by absorbing the sunlight. Simpler installations such as solar air heat collectors may be utilized. The term is also applied to solar hot water panels.

Heat exchanger – A device for transferring the heat of one substance to another, as from the solar hot water panels or the geothermal loop. Other applications may be such as a wood burning insert or a boiler.

Heating coils – Heating coils are a coil of copper tubing for example that are in direct sunlight to gather heat or in a wood burning stove for example maybe even in a boiler, they bring a source of heat from one place to another for use.

Heating tubes (on demand water heater) – These are very small tubes that wrap around within a unit such as an on demand water heater that collect heat into the copper tubing for example to heat the water while it passes through it. Another common type of heating tube is the radiant heat tubs located in a floor or wall to keep the temp in the room without air circulating and are typically located in a thermal mass such as concrete.

Hertz – Unit of frequency equal to one cycle per second. Unit used to measure the frequency of vibrations and waves such as electromagnetic waves.

High efficiency – The state of being efficient. Accomplishment or ability to accomplish a job with a minimum expenditure of time effort as well as energy.

High tension – Subject to or capable of operating under relatively high voltage:

high-tension wire. High-tension wires used to carry electrical power over long distances sustain voltages over 200,000 volts.

Hot air panels – The ultra violet rays from the sun, shine through the solar panel heating the air inside the panel. The panel reaches about 90 degrees inside.

Humanity – All human beings collectively; the human race; humankind. The quality of being humane; kindness; benevolence.

Humidity – Humid condition; moistness; dampness. An uncomfortably high amount of relative humidity.

Hydrocarbons – Any amount of compounds containing only hydrogen and carbon, as an alkane, methane, CH_4, an alkene, ethylene, C_2H_4, an alkyne, C_2H_4, along with an aromatic compound, benzene, C_6H_6. Many hydrocarbons are used as fuels are the components of gasoline; methane, which is the main ingredient of natural gas; and some components of wood.

Hydrogen – A flammable colorless gas that is the lightest and most abundant element in the universe. It occurs mainly in water and in most organic compounds. Normally consisting of one proton and one electron.

I

Imaginations – Forming mental images or concepts of what is not actually present to the senses. Synthesis of data from the sensory manifold into objects by means of the categories. Ability to face and resolve difficulties; resourcefulness.

Immersing – To plunge yourself into. To involve deeply; absorb yourself.

Implemented – To fulfill; perform; carrying out. To put into effect according to or by means of a definite plan along with a procedure.

Important – Of much or great significance or consequence: an important event in world history. Pompous; pretentious; obsolete. Mattering much.

Impose – To force on another or others. To establish something to be obeyed or complied with.

Improved – To bring into a more desirable or excellent condition. To make improvements, as by revision, addition, or change. To make more useful, profitable, or valuable.

Impure – Mixed or combined with something else. Marked by unsuitable or objectionable elements. Not pure; mixed with extraneous matter, especially of an inferior or contaminating nature: impure water.

Incandescent light (bulb) – Emitting light as a result of being heated to a high temperature; white-hot.

Inclined – To have a mental tendency, performance, ect. ; be deposed. To tend in character or in course of action. Having a disposition; tendency.

Increases – To make greater, as in number, size, strength, or quality. To multiply by propagation. The act or process of increasing. To make greater.

Independent – Not influenced or controlled by others such as in gas, water, along with electric companies. Not dependent; not depending or contingent upon someone else. Not relying on another or others for aid or help. Working for one self. Not dependent on anything or anyone else for functionally.

Indirection – A lack of direction or goal; aimlessness. Indirect action or procedure.

Induction – Electricity, magnetism the process by which a body having electric or magnetic properties produces magnetism, an electric charge. The act of causing, or bringing on, or about. The process by which an electrical conductor becomes electrified when near a charged body, by which a magnetizable body becomes magnetized when a magnetic field or in the magnetic flux set up by a magnetomotive force, also by which an electromotive force is produced in a circuit by a varying the magnetic field linked with the circuit.

Inefficient – Not efficient; unable to effect or achieve the desired result with reasonable economy of means.

Infinite – Immeasurably great. Unbounded or unlimited; boundless endless. Indefinitely or exceedingly great.

Inflationary – Of, relating to, causing, or characterized by inflation: Inflationary prices. Something costing more year after year.

Informing – To give impart knowledge of a fact or circumstance to. To supply with knowledge of a matter or subject. Inform on, to furnish incriminating evidence about.

Ingenious – Characterized by cleverness, originality of invention, or construction. Cleverly inventive or resourceful. Intelligent; showing genius. Possessing or done with ingenuity; skilful or cleaver.

Inherit – To receive by succession or will, as an heir. To receive as if by succession from predecessors. To receive property.

Insulation (Insulating) – The action of separating a conductor from conducting bodies by means of nonconductors so as to prevent transfer of electricity, or heat. Material used for insulating.

Introduced (fresh air) – To bring inside fresh air from the outside. To bring in and establish in a new place also environment.

Inventions – The creation of a new culture trait, pattern, ect. A new, useful process, machine, improvement, ect., that did not exist previously and that is recognized as the product of some unique intuition also a genius, as distinguished from ordinary mechanical skill along with craftsmanship. A device that reflects and recognized contribution to, along with the advancement of science. The act of finding.

Inventor – One who devises some new process, appliance, machine, or article; one who makes inventions.

Involuntary – Acting, done without, or against one's will. Carried out without one's conscious wishes; not voluntary; unintentional.

K

Kilowatt – A unit of power equal to 1000 watts. Abbreviation: kW, kw A measurement of power.

Knowledge – Acquaintance with facts, truths, principles, as from study or investigation. The fact or state of knowing; the perception of fact or truth; clear and certain mental apprehension. Awareness, as of a fact or circumstance.

L

Laboratory – A building, part of a building, or other place equipped to conduct scientific experiments, test, investigations, ect., or to manufacture or the like. Any place, situation, set of conditions, or the like, conductive to experimentation, investigation, observation, ect.; anything suggestive of a scientific laboratory.

LED (diode lights) – A diode light or LED light is a semiconductor light source. They are lower energy consumption, longer life and start off at only 1 watt of power consumption.

License – A certification, tag, document, ect, giving official permission to do something. Formal permission from a governmental or other constituted authority to do something, as to carry on some business or profession.

Licensed electrician – A person, who installs, operates, maintains, repairs electric devices, or electrical wiring with formal permission.

Life sustaining – To supply with food, drink, and other necessities of life such as water along with electricity for basic daily living.

Light waves – Light waves are the only electromagnetic waves visible to the human eye. These waves appear as the colors of the rainbow and each color has a different wavelength.

Lightning harvesting – Is just a dream for now, but being able to catch it like rain water along with solar power would be helpful at the least.

Lime – A calcium compound that builds up on water heaters heating tubes (in an on-demand water heater) or heating elements in an electric water heater.

Literally – Actually; without exaggeration or inaccuracy. In effect; in substance; very nearly; virtually. In a literal or strict sense.

Lobbying – A group whose members share certain goals along with working to bring about the passage, modification, or defeat of laws that affect these goals. To attempt to influence or sway towards a desired action.

Local tower supply (water tower) – A vertical pipe or tower into which water is electrically pumped to a height sufficient to maintain a desired pressure for distribution to customers. A reservoir or storage tank mounted on a tower-like structure at the summit of an area of high ground in a place where the water pressure would otherwise be inadequate for distribution at a uniform pressure.

Location – A place of settlement, or residence. A tract of land of designation. A site or position.

Low flow toilet – One that uses significantly less water than a normal one. Most use 1.6 gal and there are some that have number one flush one number two flush that use 1.6 gal, or .08 gal.

LP gas refrigerator – Propane refrigerators use heat to make cold through the natural power of evaporation and condensation. A tiny pilot light heats ammonia mixture which is permanently sealed inside the service-free cooling coils. The electricity required to operate the unit is 12 volt DC and runs on a battery bank that is recharged by a PV solar system. There are no mechanical part so no wear and tear and are highly dependable.

M

Magnetic action – Of, pertaining to a magnet, or magnetism action. Of, producing, or operated by means of magnetism.

Maintenance – The act of providing basic and necessary support. The upkeep of equipment. Necessary cleaning. The act of maintaining.

Manufacture – The making of goods by manual labor or by machinery, esp. on a large scale. To make or produce by hand. To produce in a mechanical way without inspiration or originality.

Mathematical – Having the exactness, precision, or certainty of mathematics. Of, used in, or relating to mathematics; exact. Employed in the operations of mathematics.

Mechanically – Pertaining to design, use, understanding, ect. Of tools and machinery. Belonging or pertaining to the subject matter of mechanics. Having to do with machinery.

Megahertz – One million cycles per second. Abbreviation MHz

Mentioned – A direct or incidental reference; mentioning. Formal recognition for a meritorious act or achievement. To refer to briefly to; specify, or speak of.

Mercury – A heavy, silver-white-metal, highly toxic metallic element, that is only liquid at room temperature. Atomic number is 80; melting point 38.83 deg C, boiling point 356.73 Deg C. Its chemical symbol is Hg. Also called hydrargyrum. A chemical that is gasified into the air by burning coal.

Micron – The millionth part of a meter. A colloidal particle whose diameter is between .02 and 10 microns. A unit of length equal to 10—6 metre. It is replaced by the micrometer, the equivalent SI unit.

Mindset – An attitude, disposition, or mood. An intention or inclination. A fixed mental attitude, disposition that predetermines a person's responses to and interpretations of situations.

Modified sine wave inverter – A tool that converts 12DC power to 120volt AC power with a modified sine wave form. It's an inverter that uses sine wave frequencies in-between pure sine waves and block (or square) sine waves to produce AC power from DC solar power.

Monitor – A piece of equipment that warns, checks, controls, or keeps a continuous reading of a voltage amount in the solar system. To check. A usually electronic device used to record, regulate, control a process, or system.

Monolithic crystal – A single chip of silicon. Of or pertaining to an integrated circuit formed in a single chip. A blue in color, crystal silicon chip that is in solar panels that have a typical life expectancy of 80+ years.

Motion – An act, process, or instance of changing place. The process of continual change in the physical position of an object.

Motivate – To give incentive to. To provide with a motive or motives; incite; impel.

Municipalities –A community under municipal jurisdiction. A city, town, along with other district processing corporate existence and usually its own local government.

N

National electric code – A United States standard for the safe installation of electrical wiring and equipment. It is part of the national fire codes series published by National Fire Protection Association. While it is not itself a U.S. law NEC use is commonly mandated by state or local laws, as well as many jurisdictions outside the United States.

National electric grid – A network of high-voltage power lines connecting major power stations. The three main functions of the electrical industry are the generation, transmission and distribution of energy.

Natural gas reserve – Is a gas consisting of primarily methane, typically 0-20% higher hydrocarbons (primarily ethane) and is located below ground is areas all over the world.

Net metering system – A metering system that a system owner receives retail credit for at least a portion of the electricity they generate. Net metering can be implemented solely as an accounting procedure, which requires no special metering, any prior arrangement, agreement, or notification.

Nobel Prize – A prize for outstanding contributions to chemistry, physics, physiology also medicine, literature, economics, and peace that may be awarded annually. Est., in 1901. The recipients are chosen by an international committee centered in Sweden, except for the peace prize which is awarded in Oslo by a committee of the Norwegian parliament.

Non-payment – Failure or neglect to pay: Their water, gas, and electric were shut off for non-payment.

Non-sinusoidal – Having a magnitude that does not vary as the as the sine of an independent variable: a non-sinusoidal current

Noticeably – Attracting notice or attention; capable of being noticed: a noticeable amount of interest. Worthy or deserving of notice also attention; noteworthy.

Noxious – Harmful or injurious to health or physical well-being: noxious fumes. Poisonous or harmful.

O

Observer – Someone or something that observes. A person or thing that observes.

Oceans – The vast body of water that covers almost three fourths of the earth's surface. Any of the geographical divisions of this body, commonly given as the Atlantic, Pacific, Indian, Arctic, and Antarctic oceans.

Ohms – The SI unit of electrical resistance, defined to be the electrical resistance between two points of a conductor when a constant potential difference applied between these points produces in this conductor a current of one ampere. The resistance in ohms is numerically equal to the magnitude of the potential difference.

Oil – Another name for petroleum. A thick, flammable, yellow to black mixture of gaseous, liquid, and solid hydrocarbons that occur beneath the earth's surface. It can be separated into fractions including natural gas, gasoline, naphtha, kerosene, paraffin wax, asphalt, fuel, and lubricating oils, which is used as raw material for a wide variety of derivative products. It is believed to originate from the accumulated remains of fossil plants and animals, esp. in shallow marine environments.

On demand water heaters (tank less) – Tank less or on demand water heaters are water heaters that heat the water as you use it therefore they never run out of hot water. They come in LP gas, natural gas or electric models. Like regular water heaters they need maintenance if lime is present in your water supply. They are extraordinary efficient and are much greener than many alternatives.

Operate – To bring about, effect, or produce, as by action also the exertion of force or influence. To work, use a machine, apparatus, or the like. To work, perform, or function, as a machine does.

Opposition – The action of opposing, resisting, or combating. The condition that exists when two waves of the same frequency are out of phase by one-half of a period.

Oscillations – A flow of electricity changing periodically from a maximum to a minimum. Especially: A flow periodically changing direction. A single swing or movement in one direction of an oscillating body.

Oscillators – Electronics. A circuit that produces an alternating output current of a certain frequency determined by the characteristics of the circuit components. A device or machine producing oscillations.

Oscilloscope – A device that uses a cathode-ray tube or similar instrument to depict on screen periodic changes in an electric quantity, as voltage also current. Changes in the magnitude of the potential across the plates deflect the electron beam and thus produce a trace on the screen.

Overheating – The state or condition of being overheated; excessive heat. To make too hot.

Ozone purification – The act of treating with ozone for the purpose of sterilizing water. A form of oxygen or O-3.

P

Parallel – Of or relating to two electrical wires that are separated everywhere from each other by the same distance and never connecting.

Parallel system (parallel photovoltaic system) – A photovoltaic system that consist of PV solar panels, batteries, and an inverter that supplies power to a place of use that is sometimes powered by the grid in which the power supplies never actually connect.

Perception – The ability or capacity to perceive. Insight or initiation gained by perceiving. Recognition and interpretation of sensory stimuli based chiefly on memory.

Perfect – Conforming absolutely to the description or definition of an ideal type. Exactly fitting the need in a certain situation or for a certain purpose. Accurate, exact, or correct in every detail.

Performance – The manner in which; also efficiently with which something reacts and fulfills its purpose. A particular action, deed, or proceeding. The execution or accomplishment of work.

Periodic wave – Having marked or repeated cycles. Characterized by a series of successive revolutions.

Phantom power (load) – Phantom power is power that is being consumed without knowing that it is by electronic devices that consume power simply by being plugged in while not being turned on. Noting or pertaining to a phantom circuit.

Phenomenon – A fact also an event of scientific interest susceptible of scientific description and explanation. Anything that can be perceived as an occurrence or fact by the senses. An observable fact or event.

Photoelectric effect – The emission of electrons from a material, such as a metal, as a result of being struck by photons. Some substances, such as selenium, are particularly susceptible to this effect. The photoelectric effect is used in photoelectric and solar cells to create an electric potential. Also called photoemission. The ejection of electrons from a solid by an incident beam of sufficiently energetic electromagnetic radiation.

Photovoltaic – Capable of producing a voltage, usually through photoemission, when exposed to radiant energy, especially light. Of, also concerned with, producing electric current, or voltage caused by electromagnetic radiation, esp. visible light from the Sun.

Pipes – A hollow cylinder of metal, plastic, used for the conveyance (transport) of water, gas, ect.

Planet – A large celestial body (as earth), smaller than a star but larger than an asteroid, that does not produce its own light but is illuminated by the light from a star around which it revolves.

Pollution – The contamination of air, water or soil by substances that are harmful to living organisms. The addition of any substance or form of energy (e.g., heat, sound, and radioactivity) to the environment at a rate faster than the environment can accommodate it by desperation, breakdown, recycling, or storage in some harmless form.

Possibilities – The state of being possible. Synonyms-chance, prospect, likelihood, odds. Something possible: He had exhausted every possibility but one.

Potencies – Efficacy; effectiveness; strength. Capacity to be, become, or develop; potentiality.

Potential - Electric potential. Someone or something that is considered a worthwhile possibility. Capable of being or becoming. Possible as opposed to actual.

Power loss – Energy loss due to a power outage from the grid, substation, transformer, downed power line, or lines. Causes can be from trees, tree limbs, wind, solar flares, power plant failures, cyber attacks, ect.

Power production – Power produced from climate changing sources such as: coal, natural gas power plants or dangerous power producers such as nuclear power plants. Green or environmentally friendly non climate changing power sources such as solar power, wind turbans, hydro power ect.

Power restriction – Caused by a wire that is too small for the watts and amps to flow over it efficiently. Too long of distance for power to flow over with DC current typically because of power restrictions. Improper number of wire strains for the power to flow over freely without power restrictions.

Precisely – Being exactly, strictly stated, defined, or fixed. Being exactly that and neither more or less.

Pressurized – To apply pressure to. To maintain normal water pressure in your water system.

Prime the water line – Filling the water supply line before attaching to the water pump as to eliminate any air.

Problem – Any question also matter involving doubt, uncertainty, or difficulty. Anything that is difficult to deal with, solve or overcome.

Prolonged – To lengthen out in time; extend the duration of; cause to continue longer. To make longer in spatial extent.

Propane storage – Tanks designed to store high pressure liquefied propane gas that come in sizes from 20 lb cylinders to 500 gal large tanks and larger.

PSI – Abbreviation for pounds per square inch.

Pumping – The act also process of pumping or the action of a pump. A device used to raise or transfer fluids. Most pumps function either by compression or suction or both.

Pure sine waves – The pure sine wave is an inverter's function that describes a smooth repetitive oscillation.

Pure sine wave inverters – Pure sine wave inverters are the purest of oscillations available to operate your most sensitive electronics.

PV – Abbreviation for photovoltaic.

PV array – A linked collection of photovoltaic modules, which are in turn made of multiple interconnected solar cells. The cells convert solar energy into direct current (DC) electricity.

PV power – Is a method of generating electrical power by converting solar radiation into (DC) direct current electricity using semiconductors that exhibit the photovoltaic effect.

PV system – Is a system which uses solar cells to convert light into electricity. A photovoltaic system consists of multiple components, including cells, mechanical also electrical connections also mountings with means of regulating and/or modifying output.

Q

Qualities – The character of a proposition as affirmative or negative. A personality or character trait: kindness is one of their many good qualities.

Quantity – An exact amount specified or measure. Magnitude, size, volume, area, or length.

Quest – A search, also a pursuit made in order to find or obtain something.

The act or instance of looking for also seeking; search: A quest to be solar utility independent.

R

Radiant heat – Heat that is transmitted by electromagnetic waves in contrast to heat transmitted by conduction or convection. The Sun gives off radiant heat.

Radiant heat technology – Is a technology that takes radiant heat and sends it through tubs in your floors to heat a thermal mass such as concrete to evenly spread the radiant heat throughout your place of use. The heat rises up from your floor and/or walls to give you the most comfortable climate without blowing dust around as in conventional heating applications.

Radiant heat tubes (hot water collection tubes) – Tubes that carry heat from your heat exchanger to the place of use and coil in your floors to evenly distribute the radiant heat in coils throughout the thermal mass.

Radiation – The process in which energy is emitted as particles or waves. The energy transferred by these processes.

Rain collection system – Is the accumulation and storing, of rainwater. Is a method of catching filtering and storing rainwater for later use.

Realistic – Showing awareness and acceptance of reality. Of, or pertaining to realists, also realism.

Reasonable – Agreeable to reason or sound judgment; logical. Capable of rational behavior, decision, ect.

Reduction – The amount by which something is lessoned or diminished. The act, process, result, or reducing.

Redundancy – The provision of additional or duplicate systems, equipment, ect. , that function in case an operating part fails. Duplication of components in electronic or mechanical equipment so that operation can continue following a failure of a part.

Reflective roofing – Is a roofing material capable of reflection of the heat from the sun. It blocks the radiant heat by deflecting it back up.

Regulators – A device for maintaining a designated characteristic, as voltage, current, at predetermined value, or for varying it according to a predetermined plan. A device for controlling the flow of electricity.

Remonstrating – To say also plead in protest, objection, or disapproval. To present reasons in complaint; plead in protest.

Renewable – To replenish. Relating to a natural resource, such as solar energy, water, or wood, that is never used up also that can be replaced by new growth.

Reserve – To keep back or save for future use, disposal, treatment, ect. Something stored for future use or need.

Resistance factor – A measurement of the degree to which a substance impedes the flow of electric current induced by voltage. Resistance is measured in ohms.

Resources – A source of supply, support, or aid, esp. one that can be readily drawn upon when needed. An action or measurement to which one may have resources in an emergency; expedient.

Restrict – To confine or keep within limits, as quantity. To restrain.

Restrictions – Something that restricts: as a regulation that restricts or restrains. A limitation on the use of property.

Reverse osmosis – A method of producing pure water by forcing saline also impure water through a semi permeable membrane across which salts or impurities cannot pass. Reverse osmosis is used for water filtration, for desalinization of seawater, or other impure water sources.

Revolutionize – To bring about a radical change. To inspire or infect with revolutionary ideas: such as a solar independent utility system.

Rudimentary machines – Being in the earliest stages also development of a device or invention. Of the nature of a rudiment; undeveloped or vestigial.

S

Secure – Free from also not exposed to danger or harm. Free from care; without anxiety. Safe; certain; assured.

Sediment – Solid fragmented material, such as silt, sand, gravel, chemical precipitates, and fossil fragments, that is transported also deposited by water, ice, along with wind or that accumulates through chemical precipitation also secretion by organisms, which also forms layers on the earth's surface. The matter that settles to the bottom of a liquid.

Seemingly – Apparent; appearing, whether truly or falsely, to be as specified. In appearance but not necessarily in actuality: seemingly endless hot water.

Self reliance – Reliance on oneself or one's own powers, resources, ect.

Self taught – Taught to oneself or by oneself to be (as indicated) without the aid of a formal education. Learned by oneself: a self-taught mastery of the guitar.

Sensitive – Having acute mental or emotional sensibility; aware of and responsive to the feelings of others. Quickly responsive to external influences and thus fluctuating, also tending to fluctuate.

Sensitive electronics – Transient surges from power sources can degrade these sensitive electronics as well as some types of electromagnetic radiation can interfere with them also.

Series – An end-to-end arrangement of the components, as resistors, in a circuit so that the same current floes through each component. Electricity. Consisting of or having component parts connected in series: a series circuit; a series generator; wiring 12 volt batteries in series to create a 24 volt, 48 volt battery bank; wiring two 12-16 volt variable solar panels in series to create a 24 volt current, or two 24 volt solar panels in series to create a 48 volt current for example.

Shading – To obscure, dim or darken. To produce shade in or on. A place or area of comparative darkness, as one sheltered from the Sun. A tree shading solar panels or an overhead wire that can lessen the PV production especially in a solar system wired in series as the circuit is larger is more greatly affected.

Significant – Of or relating to a difference between a result derived from a hypothesis also its observed value that is too large to be attributed to chance and that therefore tends to refute the hypothesis.

Simpler – Easy to understand, deal with, use. Not complicated; not complex. Inconsequential or rudimentary.

Sine waves – A periodic oscillation, as simple harmonic motion, having the same geometric representation. A pure (smooth) oscillation that sensitive electronics can operate without fail.

Sinusoidal – Having a magnitude that varies as the sine of an independent variable: a sinusoidal current. Having a magnitude that varies as a sine curve.

Skeptically – Not convinced that something is true; doubtful. Denying or questing; having doubt; showing doubt.

Solar array – Electrical device consisting of a large array of connected solar panels. Many solar panels in a solar system or a collection of solar panels.

Solar flares – A sudden eruption of hydrogen gas in the chromosphere of the Sun, usually associated with sunspots. Solar flares may last between several hours and several days.

Solar independent utility system – A system that operates your electricity, water, heating, cooking, refrigeration, as well as water heating, totally independent of all utilities; totally self sustaining; self reliant using only your surroundings with all the proper tools, devices, appliances and storage of necessary items mentioned in this manual.

Solar panels – A 12 volt, 24 volt, DC electric producing panel that takes direct sunlight and converts it into an 80+ year dependable electric source totally independent of any utility. Solar panel electric current can vary from 15 watts to 250 watts a panel and built into an array to be capable of producing unlimited power supplies to run any electrical device on the planet; home, business, factory, also industry.

Solar path finder – A tool that when placed in a proposed site for a solar system will revile all shaded areas to enable you to precisely calculate the total power available in that area over an entire year so you will know how much PV you will need in that location to exact a system to meet your electrical need.

Soldered – Any of various alloys fused and applied to the joint between metal objects to unite them without melting the objects to the melting point. To mend repair or patch up. To become soldered or united; fused together. To make solid.

Solution – The act of solving a problem, question. The state of being solved. The specific answer to or way of solving a problem.

Sound waves – A longitudinal wave in an elastic medium, esp. a wave producing a sound.

Source – Anything also a place from which something comes, arises, or is obtained, produced; origin.

Specifically – Relating to a specified or particular thing; Explicit.

Split ends (wires) – A wire that come to an end before it reaches its connecting point and is connected to another wire without soldering. Two ends of wires that are joined together but, are not soldered. A wire that travels over a distance of any amount but has cuts or splits along the wire that are not soldered together.

Spring water – Underground water that is held in the soil and in pervious rocks. Water from a spring.

Square footage – An area with sides of one foot in length, an example of a ten foot by ten foot square would have 100 square feet inside it. A useful calculation to define the area a solar system may be placed.

Square sine wave (block) – A kind of non-sinusoidal waveform, most typically encountered in electronics. An ideal square wave alternates regularly and instantaneously between two levels; AS a square wave on the oscilloscope.

Square sine wave inverter (block) – A inverter device that produces a square sine waveform electricity by converting 12 volt DC current into 120 volt AC current; as well as 24 volt DC current into 240 volt AC current.

Stabilizing – To maintain at a given also unfluctuating level or quantity. To make or hold stable, firm, also steadfast.

Standalone system – A system that operates independently of, also is not connected to, an electric transmission and distribution network. The standalone system is made up of a solar panel, battery bank (or reserve) and an Inverter that is capable of powering electronic devices free of any utility's.

Strategic – Important or essential to strategy. Synonyms- opportune, critical, key, principle, crucial.

Substance – That which has mass and occupies space; matter. That of which a thing consists; the physical matter of a material.

Substantial – Of ample or considerable amount, quantity, size. Of real worth, value, or effect.

Substation – An auxiliary power station where electrical current is converted, as from AC to DC current, voltage is stepped up or down. An installation at which electricity is received from one maybe more power stations for conversion from alternating to direct current, reducing the voltage, or switching before distribution by a low-tension network.

Sulfuric acid – A strong corrosive acid. It is also the acid in lead-acid electric batteries. Chemical formula H-2 SO-4 . Also called oil of vitriol.

Sun angle – The amount of sunlight a location receives is directly a result of the sun's angle. For example in the in the summer time in the northern hemisphere or in the southern hemisphere the sun's angel is more directly overhead resulting in the longest days; the longest day in the northern hemisphere is approximately June 21; and the southern hemisphere December 21.

Surge amps – Surge amps is a term to describe the extra amps a motor, television, microwave, also many other household appliances have for the start up power, but usually only for a brief period of time; this can be very taxing on your solar systems power reserve(battery bank).

Survival – The act or fact of surviving, esp. under adverse or unusual circumstances. To stay alive.

Sustainable (sustainability) – To keep up or keep on going. To provide for by furnishing means.

Switches – A device for turning on or off also redirecting an electric current along with for making also breaking a circuit. To connect, disconnect.

Switches (SCR's and TRIAC's) – SCR's and TRIAC's are high-speed; solid state switches used in AC and DC power control applications. They are sensitive electronic components that might not work well or be damaged with a squire sine wave inverter also a modified sine wave inverter, but they will operate normally with a sine wave also (pure sine wave) inverter also your grid-tie pure sine wave inverter with your solar system.

T

Totally utility independent – Totally utility independent is operating your utilities 100% off the environment also your surroundings with only storage of products, appliances, and utilizing your environment to provide you with all of your electricity, water, heating, water heating , refrigeration, along with cooking methods with absolutely no utilities whatsoever; Solar power is what makes this all possible.

Towards – In the direction of. About to become; imminent: we are moving towards a greener society with the renewable power of solar electricity.

Tracking PV array – Is a PV system that is hooked up to light sensors that follow the movement of the Sun across the sky and reset themselves to where the sun will rise in the morning automatically with motors. They can increase your power production by up to as much as 30% in a calendar year but not nearly as much in winter months due to the Sun's angle.

Transferred – The conveyance or removal of something from one place to another. A passing of something from one to another.

Transform – To increase or decrease (the voltage and current characteristics of an alternating-current circuit), as by means of a transformer. To change in form, appearance, or structure.

Transformers – An electrical device consisting essentially of two or more windings wound on the same core, which by electromagnetic induction transforms electric energy from one set of one also more circuits such that the frequency of the energy remains unchanged while the voltage along with the current usually stay the same.

Transmission lines – A system of conductors, as coaxial cable, a wave guide, or a pair of parallel wires, used to transmit signals. Wires that go from the power plants all across the country that inter connect forming what we called the Grid that consist of transmission line that carries the highest voltage to sub-stations that is transformed into a lower power source that supplies a local substation that supplies the transformer that supplies your house business or factory as well as industry.

Transmitting – To cause to spread; pass on. To cause light heat sound electricity to pass through a medium.

Transportation fleet – The vehicles that carry also transport cargo supplies goods along with people from one place to another consisting of cars trucks plans trains boats and relational vehicles.

Tremendous – Extraordinarily great in size, amount, or intensity. Extraordinary in excellence.

Trigonometric – The study of the properties and uses of trigonometric functions. The branch of mathematics that deals with the relations between the sides also angles of plane along with spherical triangles, and the calculations based on them.

Triumphs – To bring about a happy or successful conclusion (to an event, problem, ect), esp. unexpectedly: The successful completion of their solar independent utility system gave them the triumph they were looking for from utility pollution and their undependability as well as the taxing way of the utilities on the human race.

U

Ultrasensitive electronics – Electronics that are highly sensitive to electromagnetic radiation as well as the SCR's and TRIAC's switches are prone to failure.

Unique – Being the only one of a particular type; single. Having no like or equal; unparalleled; incomparable: They were unique for their solar independent utility system as the rest of the community was without utilities they were functioning absolutely normally to their neighbor's amazement.

Universe – The whole world cosmos; the totality of known or supposed objects and phenomena throughout space; the cosmos; macrocosm.

Unprotected – Not protected or unsafe from trouble, harm, unguarded: as unprotected from utility failure.

Unreliable – Not reliable; not to be relied or depended on; untrustworthy: as unreliable utilities in emergencies.

Unwisely – Not wise; foolish; imprudent; lacking in good sense or judgment: using fossil fuels to do things with was an unwisely decision.

Uplifting – Acting to raise moral, spiritual, cultural, levels: Going green gave us an uplifting feeling knowing we are doing our part.

Useable – Capable of being used: that solar power is a very usable source or power

Utilize – To put to use; turn to profitable account: to utilize the solar power was the best decision we ever could have made.

UV light radiation – Is electromagnetic radiation with a waveform shorter than the light we can see.

V

Variable temperature control (on demand water heater) – A device that compensates the temperature control for adjustment to save fuel when your water is pre-heated from another source such as solar heat.

Variety – A number of different types of things, esp. ones in the same general category: as variety of water sources.

Various – Of different kinds, though often within the same general category; diverse: as various cooking methods.

Veiled – Something that covers, separates, screens, or conceals. To hide the real nature of; mask; disguise.

Vent free radiant heater – An indoor type of heater that is 97% efficient runs of propane or natural gas and requires absolutely no electricity; has no moving parts and is highly dependable. Requires no vent.

Ventilated – To expose to the action of the wind: to vent outside.

Virtual power exchange – An exchange you have with the electric company that enables you to send power onto the grid for an electrical credit from your grid-tie solar system.

Vitally – Of or pertaining to life. Necessary to the existence, continuance, or well being of something.

Voltage (volts) – Electromotive force or potential difference expressed in volts. A measurement also the difference in electric potential between two points in space, a material, or an electric circuit, expressed in volts.

Vulnerable – Capable of being wounded or hurt. Liable or exposed: depending on utilities has left us more vulnerable than ever.

W

Water companies – A facility that collects water from many different sources depending on the area also filters treats and pumps water to water towers that distribute water to individual customers.

Water conservation – The conservation of water. Utilizing methods of saving water such as low flow toilets, overhead showers, at the tap shut off valves, pressure reduction regulators, and good habits to lessen the amount of water being used such as locating an on-demand water heater as close to the place of use as possible.

Water pumps (12 volt) – A 12volt water pump is a pump that runs off of 12 volt DC current that is produced directly from solar panels without an inverter or a battery when the sun is shining on the solar panel.

Water recycling system – A system that enables you to pump the water continuously throughout your water system that keeps it from freezing up in the coldest of freezes by drawing from a water reserve sending it through the water system then returning it to that water reserve and also gives you the ability to pump water through a 2 micron activated carbon filter to purify it.

Water restrictions – Depending upon location, these restrictions can include restrictions on watering lawns, washing vehicles, hosing in paved areas, refilling swimming pools, among many other things; they can be voluntary or involuntary.

Water storage system – A water storage system is a system of collecting rain also spring water for later use and should be degrade free as well as bug also pest proof.

Water towers – A vertical pipe or tower into which water is pumped by electricity to a height sufficient to maintain a desired pressure. A reserve or storage tank mounted on a tower-like structure at the summit of an area of high ground in a place where the water pressure would otherwise be inadequate for distribution at a uniform pressure.

Watt monitor – There is no absolute measure of electricity. Energy is defined by the work that one system does to another system. Watt, kilowatt and joule are common units of measure for electricity. There is a monitor that does monitor the kilowatts and watts at the same time to show what an electrical device is using while in use or even power being consumed when not in use called phantom power.

Wattage – Power, as measured in watts. The amount of power required to operate an electrical appliance or device. An amount of power, especially electrical power, expressed in watts or kilowatts.

Watts – The SI derived unit used to measure power, equal to one joule per second. In electricity, a watt is equal to current (in amperes) multiplied by voltage (in volts).

Well water – Underground water that is held in the soil and in pervious rock.

Wind loads – The design of a solar system must account for wind loads, and these are affected by wind shear. The pressure can be very great and if not properly mounted blow your solar system right off your roof; they must be bolted to the inside of your roof with proper bracing ; you must contact a engineer to calculate the exact size your current roof may hold.

Wires – A slender, stringlike piece also flexible metal, usually circular in section, manufactured in a great variety of diameters and metals depending on its application.

Wood stove – A space heater that is intended to heat space directly. It uses wood for fire that gives you the comfort of radiant heat.

Worldwide – Extending or spread throughout the world. Also world-wide.

Y

Yard mounted system – A yard mounted solar system is a system mounted in a yard at a specified direction towards the equator with the proper sun angle taken into account. Is fixed solar system that is set to a specific setting towards the sun that does not move and has to be manually adjusted if adjusted at all.

MSL.

2 OT wire – Is a large wire that is capable to handle DC current without power loss to a specified point (note: you must reference the national electric code book that list the proper gauge wire for the watts and amps you plan to use with it).

40 psi – The lbs. per square inch is 40 psi (less water pressure)

60 psi – The lbs. per square inch is 60 psi (normal water pressure)

60 megahertz –Think of it as pulses of power per second in AC (alternating current) ; this is the standard by which all electric companies operate all over America sending power to our home office business as well as industry wither 120 volt, 240 volt, 480 volt and so on no matter the amperes.

Index

A

AC lighting- Pg. 19
AC power supply- Pg. 19, 21, 23, 24, 25, 27, 37, 41, 47, 69, 70, 83, 103, 107
Activated carbon filter- Pg 4, 27, 29, 31, 33, 35, 37,
Alternating current- Pg 4, 27, 29, 31, 35, 37
Ambient light- Pg 27, 46
Amperes- Pg 95
Auto shut off- Pg 43, 103
Auto turn on generator- Pg 23
Automatic setback thermostats- Pg 63

B

Backup generator – Pg 23, 24
Battery acid tester- Pg 11
Battery banks- Pg 10, 11, 12, 13, 15, 16, 23, 24, 25, 26, 27, 33, 34, 47, 48, 51, 70, 71, 103, 104

C

Charge controller- Pg 11, 51, 53
Chemical refrigerant- Pg 1
Cleaning your water heater- Pg 3
Conservation- Pg 69, 70, 77, 78
Controlling lightning- Pg 67, 83, 84
Cooking methods- Pg 39, 57

D

DC current- Pg 1, 9, 16, 17, 19, 21, 22, 23, 25, 26, 37, 41, 47, 53, 69, 103, 107
DC lighting- Pg 12, 17
DC water pumps- Pg 12, 27, 28
Deep cycle batteries- Pg 12
Different kinds of batteries- Pg 12
Direct vent- Pg 3
Discharge- Pg 103
Discharge rate- Pg 12
Disconnects- Pg 103
Drain-back water valve system- Pg 4

E

Electricity- Pg 5, 67, 77, 79
Electromagnetic- Pg 87

F

Filtering water- Pg 29, 33, 35, 37
Fixed PV system- Pg 46
Fossil fuel- Pg 73
Frequency waves- Pg 83
Fuses- Pg 12, 18, 41, 53

G

Generators- Pg 81
Geothermal cooling- Pg 61, 63
Geothermal loop- Pg 7
Geothermal steam plants- Pg 84
Green technology- Pg 8, 49, 59, 67, 73, 75, 78, 79
Greenhouse gases – Pg 79
Grid-tie inverter- Pg 15, 16, 41, 103, 107
Grid-tie photovoltaic system- Pg 6, 47, 51, 71
Grid-tie system- Pg 6, 41, 47, 48

Grounding- Pg 22
Grounding rod- Pg 25
Grounding wire- Pg 25

H

Harmonics- Pg 105
Heat collector- Pg 7
Heat exchanger- Pg 7
Heating coils- Pg 1, 3, 37
Hot air panels- Pg 7
Hot water collection tubes- Pg 59
Hydrogen- Pg 55, 73

I

Independent water system- Pg 77
Inside water storage system- Pg 35
Insulation- Pg 63, 65
Inverter- Pg 1, 15, 16, 19, 23, 25, 27, 33, 41, 47, 49, 53, 54, 69, 101, 103, 105, 107, 108

K

Kerosene – Pg 81

L

LED (diode lights) - Pg 17, 19, 20, 69
Life expectancy of solar panels- Pg 10
Lightning harvesting- Pg 67
Lime- Pg 3
Low flow toilets- Pg 43
LP gas refrigerator- Pg 1, 57, 107

M

Megahertz- Pg 99
Meter that will run backwards- Pg 49
Modified sine wave inverter- Pg 15, 54, 103, 104, 105, 107

Monolithic crystal- Pg 10

N

National electric code book- Pg 10
National electric grid- Pg 5, 49, 81
Natural gas reserve- Pg 39, 55
Net metering system- Pg 49
Nobel Prize- Pg 87

O

On-demand water heater- Pg 3, 4, 43, 57
Oscillations- Pg 99, 101
Oscilloscope– Pg 104
Ozone purification– Pg 29

P

Parallel system - Pg 5, 6, 51, 52
Perpendicular electric- Pg 101
Phantom load- Pg 19, 95
Phantom power- Pg 19
Photovoltaic- Pg 45, 46, 87
Placement of solar panels- Pg 45, 46
Power losses with shading- Pg 107
Power production- Pg 48
Propane storage- Pg 57
Pure sine wave inverter- Pg 15, 53, 54, 103, 105, 107
PV array- Pg 10, 12, 25, 26, 45, 46
PV power- Pg 51, 69, 70
PV system- Pg 10, 12, 45, 46

R

Radiant heat- Pg 57, 59, 60
Radiant heat technology- Pg 57, 59, 60
Radiant heat tubes- Pg 8
Radiant heating- Pg 57
Reflective roofing material- Pg 63

Refrigeration (with propane gas and solar power) -Pg 1, 2, 57
Rain collection system- Pg 31, 32
Rain water- Pg 28, 29, 31, 32, 43
Renewable- Pg 67
Resistance factor- Pg 25
Reverse osmosis- Pg 29

S

Sensitive electronics- Pg 105
Shading affects- Pg 9, 107
Sine wave- Pg 101
Solar array- Pg 11, 46
Solar independent utility system- Pg 2, 24, 26, 28, 29, 31, 32, 34, 35, 38, 39, 42, 44, 46, 48, 49, 51, 54, 57, 60, 62, 64, 65, 70, 72, 75, 93, 95, 97, 101, 104, 108
Solar pathfinder- Pg 45
Solar power- Pg 11, 12, 47, 48, 53, 54, 84
Soldering- Pg 25, 28
Spring water collection system- Pg 33, 34
Spring water system- Pg 33, 34
Square (block) sine wave inverter- Pg 15, 54, 104, 105, 107
Stand-alone inverter- Pg 15, 53, 54
Stand-alone system- Pg 6, 12, 51, 53, 54
Surge amps- Pg 97
Switches- Pg 103, 104

T

Tracking PV array- Pg 45

U

UV light radiation- Pg 29

V

Variable temperature control flow adjustment solar water heater- Pg 4
Vent free propane radiant heater- Pg 8
Virtual power exchange- Pg 47
Visualized- Pg 101

W

Water conservation- Pg 37, 38, 43, 44
Water filtration- Pg 28, 29
Water heating- Pg 3, 4
Water recycling system- Pg 37, 38
Water storage- Pg 35
Water storage system- Pg 31, 32, 35
Water tower- Pg 77
Wind loads- Pg 45, 46
Wind loads with PV solar panels- Pg 45, 46
Wire sizing- 25
Wires- Pg 16, 25
Wiring in parallel- Pg 12
Wiring in series- Pg 12, 21, 22
Wood stove- Pg 59

Y

Yard mounted system- Pg 9, 46

MSL.

60 megahertz- Pg 23, 83